Video Recording Technology

For a complete listing of the *Artech House Audiovisual Library,*
turn to the back of this book.

Video Recording Technology

Arch C. Luther

Artech House
Boston • London

Library of Congress Cataloging-in-Publication Data
Luther, Arch C.
    Video recording technology / Arch C. Luther.
        p.   cm.
    Includes bibliographical references and index.
    ISBN 0-89006-275-7 (alk. paper)
    1. Video tape recorders.   2. Video recording.   3. Laser recording.
    I. Title.   II. Series.
    TK6655.V5L88   1999
    778.59—dc21   98-53323
                                                                          CIP

British Library Cataloguing in Publication Data
Luther, Arch C.
    Video recording technology. — (Artech House audio and video library)
    1. Video recording
    I. Title
    621.3'88'332

    ISBN 0-89006-275-7

Cover design by Lynda Fishbourne

© 1999 ARTECH HOUSE, INC.
685 Canton Street
Norwood, MA 02062

International Standard Book Number: 0-89006-275-7
Library of Congress Catalog Card Number: 98-53323

10 9 8 7 6 5 4 3 2 1

R0171528975

Contents

Preface

Video systems would not be where they are today without recording technology. The invention of video recording freed video (television, at the time) from the limitations of live production and enabled the production of creative programs of all kinds. Recording further allowed television programs to be archived for posterity without the trouble and costs of transferring them to motion picture film. This book gives a current technology overview of video recording in all its forms.

The first video recorders were the size of several desks. They were expensive, power-hungry, difficult to operate, and unreliable, but they provided the electronic storage capability that was critically needed by the nascent video industry. This made it worth the effort to use the early equipment and also spurred developments to improve the product and solve its problems. Today, all that has been achieved and video recorders can be bought for as little as US$100, packaged with a camera in a hand-held camcorder weighing less than 1 kg, and can produce recordings that are as good as a television set can display. With the emergence of high-definition television (HDTV), recorders have stepped up to the task of handling as much as 10 times more information than normal TV, while still proving to be economic, reliable, and high-performing.

Although video recording was first developed for the TV broadcasting industry, it quickly spread to many other fields, including training and education, computers, industry, the home, outer space, and essentially everywhere the need exists for capturing and storing pictures and sound.

Such broad application is possible because of the continuous advancement of the underlying technologies needed in a recorder: special materials, magnetics, optics, electronics, software, mechanics, and so on. This flow of technology shows no signs of slowing, and we can expect even more exciting recording systems and products in the future.

Probably the most significant recent advance in recorders—in fact, in video systems generally—is the transition to digital technology. Recording digitally has many advantages, including the ability to copy recorded material again and again without degrading picture or sound quality. Repeated copying (multiple generations) is needed for modern editing to create programs that are of high quality both artistically and technically. This book focuses on digital technology and describes some of the fundamentals of video and audio, as well as their application to digital recording.

This book explores recording's underlying technologies and their application to recording products at all levels: from the home to broadcasting to professional program creation. It is written for students, engineers, technicians, and technical managers both in and outside the video field. It provides an overview to assist readers in appreciating current recorder developments and application; and in making technical decisions about recorders for industry, home, or school.

Acknowledgments

Keeping up with the fast-moving video industry is a massive task for anyone attempting to write such a broad overview. It could not have been done without the help of many friends and contacts in the industry. I particularly want to mention the help of Larry Thorpe at Sony and Stan Basara at Panasonic, who both provided all the information I asked for and more.

The book also could not have been written were it not for the World Wide Web, which is an unbelievable source of all kinds of information. Recognizing that, I have added a section at the end of the bibliography that lists all the Web sites I used or know about regarding the subject of recording and its technologies.

1

Video Fundamentals

Recording in a video enables live scenes and sounds to be captured for future replay, editing, or distribution. It is an essential ingredient in nearly all video applications, from home video to broadcast program production.

The technologies involved in the design of recording equipment are diverse and include magnetics, optics, electronic circuits, precision mechanical components, digital computers, physical packaging, and ergonomics. All of these disciplines must be carefully integrated in a successful recording product and each is a challenge, even to specialists in the field. Few other products require such a wide range of technology.

This chapter provides some background on video systems as they affect recorders.

1.1 VIDEO SYSTEMS

Electronic video always involves a *system* consisting of several units. The simplest system is a camera connected directly to a display, but such a system can operate only in real time because there is no capability to store video for later replay or editing. Most practical systems include one or more recording devices, as shown in Figure 1.1. A signal source, usually a video camera, connects to a recorder where the video is stored on tape or disk for later use. When replayed, the recorded video may be processed for enhancement, edited to create programs, or reproduced for distribution. The final video is distributed to viewers by broadcasting, cable, or recorded media.

Recorders in a video system provide storage of incoming video signals from cameras or other sources. Subsequent use of the stored video may involve replay and rerecording for editing purposes. Further rerecording of the edited video may be required to make copies for the distribution of the completed program to other users. Thus, the passage of signals through a recorder may occur several times before the completed program reaches its end users. Each pass of a signal through a recorder is called a *generation*; the editing and distribution scenario just described is an example of *multigeneration* recording.

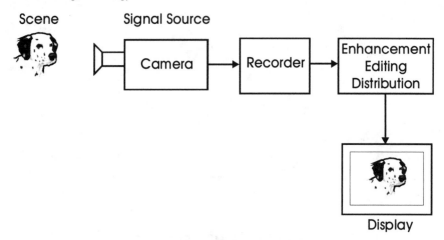

Figure 1.1 *A typical video system.*

In analog systems, multigeneration recording places demanding requirements on recorder signal performance because distortions in successive analog recorders will accumulate to produce increasing signal degradation. That is generally not the case for digital recorders, which can be designed to have essentially perfect reproduction from generation to generation. This is one reason why recorders were one of the first applications of digital technology in video and audio.

1.2 RECORDING REQUIREMENTS

Recorders are basically signal-handling devices that should handle their signals with little or no degradation, meaning that they should be *transparent* to the video or audio signal. Most video recorders are critically dependent on the format of the input signal (see Section 1.3), which must meet the signal standard for which the recorder is designed. (Certain digital recorders are not necessarily dependent on the signal format; digital signals can be recorded and replayed without knowing the internal format of the data, but this limits some features of the recorder, such as variable-speed playback or editing. See Chapter 7.)

The following list shows some of the more important requirements placed on video and audio recorders by application in different markets. This is summarized in Table 1.1.

- *Picture and sound quality*—This is limited by the signal standards chosen, but it may deliberately be further limited by the recording format. For example, the VHS format for home recorders uses standard TV signal formats but restricts the video bandwidth below that of most other standard TV devices. Production and postproduction require the highest picture and sound quality to allow for repeated processing and possible conversion to other video or audio formats, so recorders for this use strive to be as transparent as possible.
- *Preferred configuration*—The familiar camcorder configuration is used in all markets except postproduction. Standalone recorder boxes are also used in all mar-

Table 1.1
Recorder Requirements for Different Markets

Attribute	Home	Semiprofessional	Production	Postproduction
Picture quality	Standard TV or less	Standard TV	Highest possible	Highest possible
Preferred configuration	Camcorder or set-top box	Camcorder and standalone	Camcorder and standalone	Standalone
Mounting	Hand-held	Hand-held or desktop	Hand-held or rack-mounted	Rack-mounted
Editing	Built-in	Built-in	Assemble only	No
Variable-speed	Optional	Yes	Yes	Yes
Remote control	Optional	Yes	Yes	Yes

kets. Home standalone recorders are often available in a *set-top* configuration for use with a standard TV receiver.

- *Mounting*—Camcorders are usually hand-held or shoulder-held for casual shooting and mounted on tripods or pedestals for more serious applications. Semiprofessional or postproduction applications usually call for desktop or rack-mounted recorder packages.

- *Editing*—This is the control means for selection of prerecorded segments and assembly of the segments into a continuous program. Home camcorders have built-in editing capabilities because the home user generally cannot afford dedicated equipment for editing. Semiprofessional or professional applications generally have separate editing facilities, so the built-in capability of camcorders is used less frequently.

- *Variable-speed playback*—This is provided as an option in home recorders primarily for its use as a special effect during playback. In other markets, variable-speed playback is important for use during editing.

- *Remote control*—This is a desirable feature in all recorders; it becomes mandatory when the recorder is used in a system that has an external edit controller.

1.3 VIDEO SIGNAL PROPERTIES

Video signals are designed to meet certain standards in different places around the world. The different types of video are often referred to as *formats*. Because recorders are closely tied to the signals that they handle, it is important to understand them and how they affect recorders. This section covers some of the factors that govern signal properties and formats.

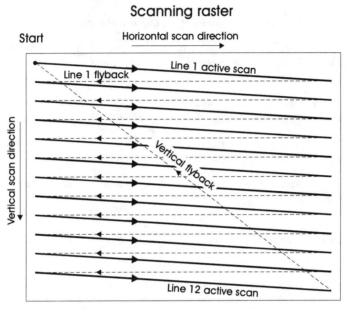

Figure 1.2 *Progressive scanning. Reproduced with permission from [1].*

1.3.1 Scanning

Video systems create an electrical signal from a two-dimensional picture by the process of *scanning*, which is done by imaging devices in a camera [1]. The brightnesses of points in a picture are reproduced in the imaging device by beginning at the top left of the picture and reading points horizontally across the picture. The sensor outputs a continuous signal that represents the variation of brightness seen as the reading moves across the picture. At the right side of the picture, scanning moves back to the left and then down to read the next line of points. This continues until the bottom of the picture is reached. Then, the scanning process repeats by going to the top left and starting over. Each complete scan of all the points in the picture is a *frame*; frames are scanned rapidly enough to allow smooth motion in the picture to be reproduced. This requires frame rates of at least 25 to 30 Hz. The actual scanning pattern is called a *raster*.

1.3.1.1 Progressive and Interlaced Scanning

Scanning all the points of the picture in a single vertical scan as just described is called *progressive scanning*, which is shown in Figure 1.2. An alternative scanning method that scans only half of the lines in each vertical scan is *interlaced scanning*, shown in Figure 1.3. Interlaced scanning is used in standard TV systems because it allows a 2:1 reduction of the frame rate, and thus the bandwidth (see Section 1.3.1.3), compared to progressive scanning. However, it does have some limitations, especially in the reproduction of near-

Scanning raster

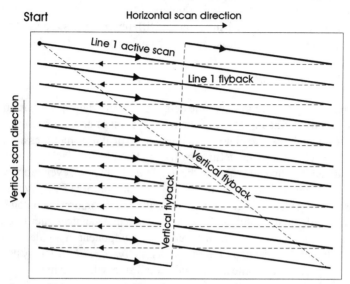

Figure 1.3 *Interlaced scanning. Reproduced with permission from [1].*

horizontal edges in the picture or in fine detail, such as text. In both of these cases, edges tend to flicker. As a result, interlaced scanning is not used in computer displays, and the new HDTV standards provide options for either progressive or interlaced scanning.

In interlaced scanning, each vertical scan includes only half of the total lines. The vertical scans are called *fields*. The first field, which starts with a full line, scans the odd lines of the raster, while the second field, starting with a half-line, scans the even lines. Thus, the field rate is twice the frame rate, which is why interlacing reduces flicker for a given frame rate.

1.3.1.2 Blanking Intervals

It is customary in video signals to allocate a certain amount of time at the end of each line and each field or frame for the scanning device to start the next scan. These times are called *blanking intervals*, horizontal and vertical. During blanking intervals, any signal from scanning is blanked out and other information, such as synchronization, may be added to the signal.

1.3.1.3 Aspect Ratio, Lines, and Frames

The proportions of a raster are defined by its *aspect ratio*, which is the ratio of raster width to raster height and is usually given in the form *w:h*, where the fraction is reduced to its smallest numbers. The aspect ratio for most analog systems is 4:3.

Other specifications of a raster are the total scanning lines, the frame rate, and the interlace ratio. For example, the scanning numbers for standard TV in the United States are 525 lines, 30 frames/s, and 2:1 interlacing. Different numbers are used elsewhere in the world and in the digital broadcasting systems now emerging (see Section 1.6).

1.3.1.4   Resolution and Bandwidth

A video system's ability to reproduce fine detail is quantified by specification of *resolution*. This is usually defined in terms of the reproduction of patterns of alternating black and white lines, which may be either horizontal or vertical or sometimes at other angles. By convention, such patterns are defined by the total number of black and white lines that would be included in a distance equal to the height of the picture, even though the pattern may not be that large. This is called the television line number (TVL). Various test patterns and test signal generators are used for evaluation of resolution in analog systems [2].

For a given horizontal resolution, faster scanning will generate higher frequency components in the video signal. The highest frequency component in a video signal may be calculated as shown in [3]; this is referred to as the *bandwidth* of the system. Because there are always parts of the scene that contain detail at less than the maximum resolution, most video signals contain lower frequency components, down to very low frequencies generated by the field or frame scanning. This is the *spectrum* of the video, which is explained further in [4].

1.3.2   Color Reproduction

Monochrome video systems generate a single signal from scanning that represents the brightness variations of the scene. This signal is called *luminance*. According to the *trichromatic system*, color reproduction requires the generation of three signals. In a color system, three channels having different responses to colors are used. The three channels are defined by their dominant colors, called *primary colors*. The outputs of the three channels are suitably combined at the display to produce the color reproduction.

1.3.2.1   Additive and Subtractive Color

Color systems may display their reproduction by either placing colored dyes on white paper, as in color printing; or they may display by adding up the output of colored light sources, as in video. The latter approach is called *additive color* because of the addition of light sources, and the former is *subtractive color* because the dyes subtract colors from the white light reflected by the white paper.

1.3.2.2   Primary Colors

The most common set of additive color primaries are red, green, and blue (RGB). Figure 1.4 shows these colors and the colors created by combining each two of the

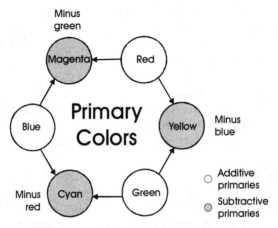

Figure 1.4 *The RGB additive primary colors and their complements. Reproduced with permission from [1].*

additive primaries, which are called *complementary colors*. Other colors are produced by different combinations of the primaries. A corresponding subtractive color system is made by using the additive complements as the primary colors. Thus, cyan, magenta, and yellow (CMY) are subtractive primaries. In color printing, better reproduction is obtained by using a fourth color, black, which otherwise would have to be made by combining large amounts of all three primaries. The system using cyan, magenta, yellow, and black is known as *CMYK*.

1.3.3 Component Video Signals

Color video originates in a camera as three signals representing the red, green, and blue sensors. These signals are called *components,* and a system that processes the three signals separately is a *component video* system.

It is useful to process the RGB components into a different set of components that has advantages in signal handling. Because color reproduction is possible with any set of signals that represent uniquely different color responses, RGB signals can be transformed to other component sets by linearly combining them in different proportions. The most common form of this is the *luminance and color-difference* structure.

A luminance signal (Y) is formed by combining RGB signals according to:

$$Y = 0.59\,G + 0.30\,R + 0.11\,B \tag{1.1}$$

This produces a signal representative of the scene's subjective brightness when displayed as a monochrome image. Two other component signals can be produced by subtracting the Y signal from R and B:

$$C_R = R - Y \tag{1.2}$$
$$C_B = B - Y \tag{1.3}$$

These are called *color-difference* signals and are characterized by becoming zero for monochrome areas of the scene. This is called the YC_RC_B component system.

Color difference signals, which appear only in colored areas of the image, may be transmitted and displayed with somewhat less resolution than the luminance signal. This is because the normal human eye has less acuity for colors than for luminance, a feature exploited in nearly all video systems.

Component video systems require three transmission channels for the three components, which is awkward and expensive for many purposes. However, methods are available to combine components into a single *composite* signal.

1.3.4 Composite Video Signals

Analog color component signals may be combined by any of the well-known multiplexing techniques, such as time-division or frequency-division multiplexing. However, further advantages are obtained by exploiting the unique properties of video signals.

1.3.4.1 Suppressed-Carrier Quadrature Modulation

Two color-diffence components may be amplitude modulated on a single carrier by the use of *quadrature modulation*. In this method, two phases (0° and 90°) of the same carrier are modulated by the two signals. On reception, the signals are reconstructed without interference by detecting the two quadrature phases of the carrier. Thus, two noninterfering signals are transmitted in the same bandwidth.

Color-difference signals are *bipolar*, meaning they can go positively or negatively from zero. Zero, of course, represents no color in the picture. Modulating with this bipolar signal causes the carrier amplitude to go to zero for a monochrome picture, and the carrier phase reverses when the color-difference signal reverses. A carrier modulated in this way with color-difference signals is called a *color subcarrier*.

1.3.4.2 Frequency Interleaving

The spectrum of a scanned video signal contains peaks of energy at the horizontal-scan frequency and its harmonics. Between these peaks, there are regions of low energy. If the spectrum of a second signal is arranged so that its peaks of energy fall into the low-energy regions of the first signal, they can be combined with minimal interference. This can be done with the luminance as the first signal and the modulated color-difference carrier as the second signal by suitably choosing the color carrier frequency so that its sidebands interleave between the sidebands of the luminance signal. Specifically, the color carrier frequency must be an odd multiple of half the line-scanning frequency.

1.3.4.3 Visibility of Interference in Luminance

A given amount of interference to a luminance video signal becomes less visible as the frequency of the interfering signal increases. Thus, interleaving a color carrier near the top end of the luminance bandwidth will give minimal interference to the luminance. This and the two features discussed above form the basis of all analog composite video systems. These systems are discussed specifically in Section 1.6.1.

1.4 ANALOG VIDEO

The properties of analog video have already been mentioned, and they are summarized here. Scanning creates a continuous signal that can have any value at any time depending on the brightness that reaches the sensor device. This is an analog signal. The continuous nature of the signal means that it is subject to amplitude or time distortions as it is transmitted or processed because any value at any time is valid so transmission distortion cannot be distinguished from the correct signal. Thus, analog signals are subject to many kinds of distortion and those distortions may accumulate as the signals are processed through the various units of a large system.

1.4.1 Analog Video Performance Measures

The performance of any system may be evaluated either by analyzing the output signals from the system to look for distortions, or by examining the system itself to see what distortions it produces. The former may be called *picture analysis* and the latter may be called *channel analysis*. Both types of evaluation are useful for analog video.

1.4.1.1 Picture Analysis

To make possible the separation of analog distortions from the picture, it is necessary to use special test pictures (patterns) whose properties are known. The output from the tested system can be viewed either as a picture or as a waveform to see how the system has changed the known properties of the pattern. This is done easily and test patterns such as resolution charts, multiburst charts, gray scale charts, color bar patterns, and grating patterns are widely used in analog systems [5].

The analog performance attributes that are amenable to picture analysis are as follows:

- Resolution;
- Transient response;
- Noise and interference;
- Gray scale (amplitude) linearity;
- Scanning (positional) linearity;
- Color balance.

There are many kinds of analog distortion and each must be tested.

1.4.1.2 Channel Analysis

Picture analysis is not capable of detecting very small distortions because of the masking effect of the video signal itself. To test system components used in large systems where the distortion allowed in a single component may be too small to see in a picture analysis test when testing that component by itself, channel analysis is used.

In channel analysis, a special test signal is used for each distortion measurement; these signals may not even be pictures at all, although they must have the appropriate characteristics, such as blanking intervals, for passing through video systems or modules. A special analyzer unit is used to evaluate the signal output to get quantitative values for each type of distortion. Channel analysis of analog video systems or modules can measure the following:

- Frequency response (bandwidth);
- Transient response;
- Phase distortion (color subcarrier transmission properties);
- Amplitude linearity (gray scale);
- Signal-to-noise ratio (SNR).

Channel measurements can detect distortions far smaller than can picture analysis and is therefore the only way to evaluate modules that may be cascaded many times in the signal path of a large system. Analog recorders for use in large systems must be tested this way.

1.5 DIGITAL VIDEO

Whereas analog systems must reproduce a continuous scale of amplitude values, digital systems use a limited number of significant amplitude values—as few as two. Systems using only two signal values are known as *binary*. Other digital systems may use more values (sometimes called *levels*), but they all share the benefit that small distortions in transmission or processing can be removed by restoring the signal to its specific levels, a process called *requantizing*. As long as the distortion does not cause the level to shift all the way from one level to the next, requantizing will restore a perfect signal. This has many advantages compared to analog technology.

- The precision of digitally representing analog quantities is established at the point of conversion from analog to digital. Once established, the digital precision can be maintained (if desired) through an extended system by the use of digital error protection. This is a perfect reproduction system (i.e., transparent).
- Digital signals can be stored in memory devices. Many processes are facilitated by storage, such as image enhancement based on storing adjacent data from the image, or conversion between different video standards.
- Complex digital functions are produced inexpensively by designing them into integrated circuits. This allows extremely sophisticated signal processing techniques to be achieved easily and economically.

- The complex process of video or audio data compression allows digital transmission or recording to be done with less bandwidth than an equivalent analog system.
- Digital systems are less expensive than equivalent analog systems, and the cost will continue to fall as integrated circuit technology becomes cheaper.

The principal disadvantage of digital technology is that digital circuits are inherently more complex than analog circuits. This originally made digital technology more expensive, but that has been overcome by the use of integrated circuits. As a result, nearly all video and audio systems, from home systems to the highest-level broadcast and professional systems, are becoming all-digital.

1.5.1 Analog-to-Digital Conversion

Nearly all video and audio signals originate as analog and must go through a conversion process at the input to a digital system. This is called *analog-to-digital conversion* (ADC). Because display devices are also analog, *digital-to-analog conversion* (DAC) must occur at the end of a digital system or at any monitoring points within the system.

The video and audio performance of a digital system is determined (and limited) at the point of conversion. Therefore, the conversion process is extremely important. Conversion requires three operations, which may not always be done in the order presented here. They are sampling, quantizing, and encoding (all are shown in Figure 1.5).

1.5.1.1 Sampling

The horizontal resolution of a digital video system or the bandwidth of a digital audio system is determined by the *sampling* process. Sampling converts an analog continuous time scale into a series of discrete points in time. A sampling circuit reads the instantaneous value of an analog signal at discrete points determined by a sampling clock of a given frequency, called the *sampling rate*. The sampled values are still analog values, but the time scale has been made discrete (digital).

The sampling rate determines the highest signal frequency that can be correctly sampled according to the *Nyquist criterion*, which states that the sampling rate must be at least twice the highest signal frequency to be sampled. This also may be stated from the signal frequency point of view: the highest signal frequency should be no greater than half the sampling frequency. This is called the Nyquist *limit*; if it is exceeded, spurious signal components, called *aliasing*, will be generated. To prevent this, most ADCs include a prefilter as shown in Figure 1.5 to remove signal frequency components that are too high.

1.5.1.2 Quantizing

Quantizing converts the continuous analog amplitude scale to digital values. A set of levels is established, as shown in Figure 1.5, and all signal values that fall between two levels are forced to have the same value. The number of levels depends on how many

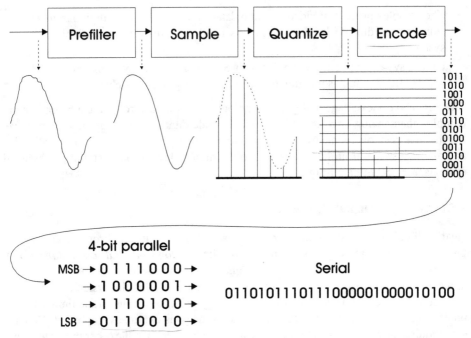

Figure 1.5 Block diagram of an ADC. Reproduced with permission from [1].

digital bits are assigned to each sample value; if the number of bits/sample is N, the number of levels is 2^N. For example, 8 bits/sample will have 256 levels. Digital video systems usually have 8 or 10 bits/sample, and digital audio systems typically use 16 or 20 bits/sample.

The *data rate* generated by an ADC is the product of its bits/sample and the sampling rate. The bits/sample is fixed by the ADC design, but the sampling rate may be varied up to a maximum determined by the ADC specification.

The principal artifact generated by quantizing is a signal-dependent error produced by the round off of analog values to fit the levels of the digital signal. It is called quantizing *error* or quantizing *noise*.[1] The visual effect of quantizing error is called *contouring* because of its similarity to the appearance of a contour map. Because contouring is signal-related, it is visible even at quite low levels. The visibility may be reduced by adding a small amount of random noise before quantizing, a process called *dithering*, which makes the quantizing error look more like noise. Most ADCs use dithering.

---

1   Note that quantizing error is not true random noise because it is correlated with the signal. However, it is customary to equate it to noise because, like noise, it is a spurious component added to the digital signal.

Quantizing determines the digital system's maximum signal-to-noise ratio (SNR), which can be calculated from the number of bits/sample as:

$$SNR \ (dB) = 6.02 \ N + 10.8 \tag{1.4}$$

Thus, 10-bit quantizing has a maximum SNR of 71 dB.

1.5.1.3 Encoding

The final step of ADC is to assign a digital bit pattern to each quantizing level, a process called encoding. Figure 1.5 shows an example of this where the levels are encoded according to a simple binary progression for the 4 bits of quantizing used in the figure. Many other encoding schemes are possible—the only caveat is that someone receiving a bit stream must know the encoding used in it. That is part of the standardization of bit streams.

1.5.1.4 Pixels

When a series of samples represents a video signal, each sample can be considered to represent a single point of the picture—a picture element, or *pixel*. Each pixel can have a brightness value and color values independent of adjacent pixels. The pixel concept is also sometimes used with analog systems by considering each half-cycle of the highest video frequency to be a pixel.

The number of pixels in a digital image determines the maximum resolution of that image. For example, a common digital image format has 640 (H) × 480 (V) pixels. With a 4:3 aspect ratio, that corresponds (approximately) to an analog image with a resolution of 480 TVL in both directions.

1.5.1.5 Digital-to-Analog Conversion

Conversion from digital back to analog (DAC) is a simpler process than ADC; it involves decoding the sample values back to the quantized levels and low-pass filtering to remove any transients produced by the process. Because part of this is analog processing, DAC is subject to some analog distortions. The ultimate performance of a digital system cannot exceed the performance of its ADC and DAC connected directly together. If the digital processing of the rest of the system is error-free, the total system will have the same performance as the ADC and DAC.

1.5.1.6 Data Compression

One of the important features of digital audio or video systems is that the data can sometimes be compressed to reduce the data rate of the system to below that of the ADC. This can be done in either *lossless* or *lossy* ways. Lossless means that no information is lost in the compression; the output after decompression is exactly the same as the input before compression.

The degree of compression available with lossless compression is usually less than 2:1 with video or audio. To achieve greater compression, many systems use lossy compression. However, lossy compression does change the signal—the trick is to only change the signal in ways that the viewer or listener will probably not notice. Using lossy methods, audio compression up to 8:1 or more is possible and video compression up to 100:1 may be achieved. Compression is covered further in Chapter 7.

1.5.2 Digital Video Performance Measures

The sound or picture performance of a digital system is determined by the ADC and DAC. These may be tested together using conventional analog methods (see Section 1.4.1). Then, if the rest of the digital system does not involve compression, it may be tested by evaluating channel error performance.[2] If the error levels are below the threshold of the system's error protection capability, the digital system is transparent. If not, the system performance will probably fail catastrophically. Error protection is discussed further in Section 7.7.

The performance of a binary digital channel is expressed in terms of data (bit) rate and *bit error ratio* (BER). The BER expresses the average number of errors in a given number of bits, expressed as a ratio. For example, if there is an average of 1 error in 1,000,000 bits, the BER is 10^{-6}. Error protection schemes are capable of improving the BER of the raw channel by a large factor, but they are limited in how many errors can be corrected and how those errors are arranged in time. The latter consideration is expressed by stating the statistics of errors being grouped together. Groups of errors are called *burst errors,* and error protection methods are limited in this regard. When burst errors are produced in a channel by such effects as dropouts on magnetic tape, additional processing can be added to break up the bursts so that error protection processing will be more effective (see Section 7.7).

1.6 VIDEO SIGNAL STANDARDS

Recorders and other signal-handling devices expect the signals they receive to meet certain requirements. Otherwise, they cannot work properly. This is the reason for standards. Without them, systems such as TV with hundreds of millions of users could not exist. This section describes the important standards in use around the world, both analog and digital.

1.6.1 Analog Video Standards

Although the video world is going digital, the analog video world has existed for more

---

2   Systems involving compression generally have to be tested through an ADC and DAC using analog methods.

than 40 years and it will exist many more years, so it cannot be ignored. There are hundreds of millions of analog TV receivers, recorders, and cameras in use that will have to interface with the digital equipment now coming out. Existing analog video standards [6] are composite; analog component systems are rare. Where components are used, the three signals are handled separately and generally follow the standards for the luminance part of the corresponding composite system.

1.6.1.1 National Television Systems Committee (NTSC)

The first composite color video system was developed in the United States in the early 1950s. It was called *NTSC*, an acronym for the body that wrote the first standards—the National Television Systems Committee. NTSC color was designed to operate in the same 6-MHz broadcast channels that were previously broadcasting only monochrome video.

A further objective of NTSC was that it had to be compatible with the monochrome system; that is, existing monochrome receivers would be able to view the color signals in monochrome. That objective required the same scanning format to be used for color. Color is added by a color subcarrier that is constructed according to the principles described in Section 1.3.4. Some specifics of the NTSC standard are shown in Figure 1.6. It is described in detail below because many NTSC features are used by the other composite video systems.

Figure 1.6(a) shows the video waveform viewed at the horizontal scanning rate. The full horizontal period is referred to as H, which is 63.5 µs. Each horizontal period is divided between the horizontal blanking interval and *active line time*, which is the time for display of picture information. From the numbers in Figure 1.6(c), it can be seen that horizontal blanking represents nearly 17% of H. This time provides for retrace of the scanning beam in displays and for transmission of synchronizing information. The latter consists of the horizontal sync pulse, which is approximately 4.8 µs in duration; and the color synchronizing signal, which is a burst of at least eight cycles of the color subcarrier frequency, having a precise phase relationship to the color modulation of the subcarrier.

Figure 1.6(b) shows details of the vertical blanking interval, which has a duration of 20H or 8% of the total vertical period. Thus, between horizontal and vertical blanking, nearly 25% of the total transmission time is devoted to blanking intervals during which no picture information is being transmitted. This was necessary in the days when the NTSC standards were established, but with today's technology this time can be considerably shortened; that has been done in the new digital standards.

It is necessary to maintain horizontal synchronization during the vertical blanking interval, so sync pulses continue during that time. A vertical sync pulse is also added, 3H in duration. To maintain horizontal synchronization during this interval, vertical sync is serrated to provide horizontal sync edges during the wide pulse. Since interlaced scanning changes the relationship between horizontal and vertical sync pulses on odd and even fields (odd fields begin on an even line but even fields begin on a half-line), the vertical sync interval has pulses and serrations at twice the horizontal frequency; this makes that interval identical for odd and even fields.

(a) Line-frequency display

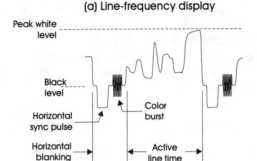

Peak white level

Black level

Color burst

Horizontal sync pulse

Horizontal blanking

Active line time

(b) Field-frequency display
(expanded to show VBI)

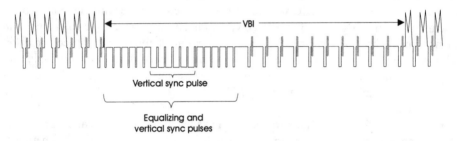

VBI

Vertical sync pulse

Equalizing and vertical sync pulses

(c) NTSC specifications

Item	Specification	Item	Specification
Total lines	525	Chrominance bandwidth (MHz)	$I = 1.3$
Interlace	2:1		$Q = 0.5$
Field rate (Hz)	59.94	Active line time (µs)	52.7
Frame rate (Hz)	29.97	Horizontal blanking interval (µs)	10.8
Line rate (Hz)	15,734.26	Horizontal sync width (µs)	4.8
Luminance bandwidth (MHz)	4.2	Horizontal front porch (µs)	1.5
Subcarrier frequency (Hz)	3,579,545±10	Color burst width (cycles of SC)	9
Subcarrier/line ratio	455/2	Vertical blanking interval (H)	20

Figure 1.6 Video signal waveforms and specifications of the NTSC composite system.

To achieve proper frequency interleaving, as described in Section 1.3.4.2, the color subcarrier frequency (SC) must be exactly synchronized with the scanning frequencies. For NTSC, the relationship between subcarrier and horizontal frequencies is 455/2. The subcarrier itself is precisely controlled at 3,579,545 Hz ±10 Hz. Most hardware begins with an oscillator at subcarrier frequency, or a multiple, and has frequency dividers to generate the scanning frequencies.

The subcarrier frequency chosen for NTSC (3,579,545 Hz) places the subcarrier close to the top of the 4.2-MHz bandwidth, with the result that the upper sideband of the modulated chrominance is cut off around 0.5 MHz above the subcarrier. In quadrature

Table 1.2

Specifications for the PAL Composite Video System

Item	Specification	Item	Specification
Total lines	625	Chrominance bandwidth (MHz)	$U = 1.3$
Interlace	2:1		$V = 1.3$
Field rate (Hz)	50.0	Active line time (µs)	53.2
Frame rate (Hz)	25.0	Horizontal blanking interval (µs)	10.8
Line rate (Hz)	15,625.	Horizontal sync width (µs)	4.8
Luminance bandwidth (MHz)	5.0 or 5.5	Horizontal front porch (µs)	1.5
Subcarrier frequency (Hz)	4,433,619±10	Color burst width (cycles of SC)	9
Subcarrier/line ratio	1135/4 + 25	Vertical blanking interval (H)	21

modulation, crosstalk between the two signals will occur if one of the modulation side-bands is removed. To prevent this crosstalk, the bandwidth of one color component is restricted to 0.5 MHz, so above that modulating frequency, there is no quadrature modulation. In NTSC, the two color-difference components are called I, for in-phase, and Q, for quadrature. Thus, the NTSC components are Y, I, and Q. The Q bandwidth is reduced to 0.5 MHz but the I bandwidth is 1.3 MHz. The color-differencing equations for I and Q are chosen to place Q in a range of colors (blues) where the narrow bandwidth is least visible to the eye.

1.6.1.2 Phase-Alternation Line (PAL)

European countries used different scanning numbers than the United States in monochrome TV, and they added color to their TV system at a later time. Because of the fundamental difference in scanning, they could not simply copy the NTSC standards. By the time European color standards were adopted, there had been considerable experience with NTSC, and some of its recognized limitations could be solved in a new system. The European standard, called PAL for *phase-alternation line*, followed the basic concepts of NTSC, but the manner of color modulation was modified to eliminate the need for narrowing one of the color-difference bandwidths and to make the system more immune to analog distortions that affected color rendition. The specifications for PAL are given in Table 1.2.

The PAL color components are called U and V; U is derived from $B - Y$, and V is derived from $R - Y$. The PAL system is named for its different color modulation system—phase-alternation line. This means that the phase of the V color subcarrier components is reversed on adjacent scanning lines. When the picture is viewed normally, PAL provides cancellation between some of the color distortions that plague NTSC. This happens because the carrier phase reversal causes distortions to be reversed in phase on alternate lines, and thus they visually cancel.

Because the color carrier phase reversal would interfere with the frequency interleaving of the V components, a further modification is required in choosing the color subcarrier

Table 1.3

Specifications for the SECAM Composite Video System

Item	Specification	Item	Specification
Total lines	625	Chrominance bandwidth (MHz)	$D_R = 1.3$
Interlace	2:1		$D_B = 1.3$
Field rate (Hz)	50.0	Active line time (μs)	52
Frame rate (Hz)	25.0	Horizontal blanking interval (μs)	12
Line rate (Hz)	15,625	Horizontal sync width (μs)	4.8
Luminance bandwidth (MHz)	6.0	Horizontal front porch (μs)	1.5
Subcarrier frequencies (Hz)	4,406,250;	Color burst width (cycles of SC)	9
	4,250,000	Vertical blanking interval (H)	25

frequency—the frequency must be offset by one-quarter line plus 25 Hz (the frame rate). This is a slight complication for PAL frequency generation.

1.6.1.3 SECAM

When PAL was being developed in Europe, France chose to develop a different system using frequency modulation of two color subcarriers. This system, which was also adopted in the countries of the former Soviet Union, is called *Sequential Couleur avec Mémoire* (SECAM).

The SECAM color-difference components D_R and D_B are also derived from $R - Y$ and $B - Y$; they are frequency-modulated on two different carrier frequencies that appear on alternate scanning lines. Thus, each color-difference signal is transmitted on only half of the lines. At the receiver, one-line memories are used to allow display of both color components on all lines. This is the reason for the word *mémoire* in the name. The alternate-line approach cuts the vertical color resolution in half, which is not normally very visible to the viewer. The specifications for SECAM are given in Table 1.3.

1.6.1.4 MAC

Another composite system that is less widely used than the three described above, is the *multiplexed analog components* (MAC) system. It has been used in satellite broadcasting and some recording applications and is interesting because of its very different approach: it uses time-division multiplexing rather than frequency-division multiplexing. A block diagram and line-frequency waveform for a MAC system are shown in Figure 1.7.

By transmitting the components sequentially in each line, crosstalk between them is avoided. Also, the effect of transmission distortions on color performance inherent in the other composite systems is minimized. To fit the components into the active line time, they must be time compressed, a process that is easily accomplished with digital circuits. However, the transmitted video signal is not digital, except for the horizontal blanking

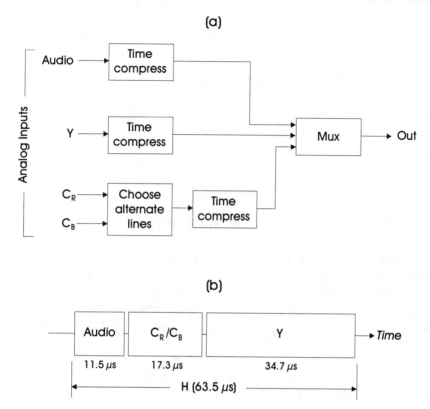

Figure 1.7 *(a) Block diagram and (b) line-frequency waveform of a MAC system.*

interval, which is used to transmit sync and audio digitally.

In one version of MAC, B-MAC, the luminance signal is time compressed in the ratio of 2:3 and the color-difference components are compressed 1:3. The color components are transmitted on alternate lines, as in SECAM, so the receiver has to store the color components to display both of them on every line. Of course, this cuts the vertical color resolution in half.

The MAC signal requires a slightly wider bandwidth (compensating for the degree of time compression of the luminance component) for equivalent performance. This is one reason that MAC is not suitable for broadcasting on existing TV channels. The other reason is that the signal is not compatible with monochrome TV.

1.6.2 Digital Video Standards

Standards for digital video may be considered at four levels.

1. *Signal origination*—These standards define scanning and digital sampling formats that may exist within cameras or other signal origination units. They are not

Table 1.4

Sampling Parameters for Component Systems Based on ITU-R Rec. BT.601-5

Standard	f_S (MHz)	Bits/Sample	Data rate (Mbps)
4:4:4	13.5	8	324
4:2:2	13.5/6.75	8	216
4:4:4	13.5	10	405
4:2:2	13.5/6.75	10	270
4:1:1	13.5/3.375	8	162
4:2:0	13.5/6.75*	8	162

* Color-difference components are also subsampled 2:1 vertically.
Reproduced with permission from [1].

complete standards for connections between physical units. Signal origination formats may also be called *picture formats*.

2. *Bit streams*—These standards describe the structure of a bit stream and the processing necessary to create it. Video or audio compression, when used, is part of a bit stream standard.

3. *Signal interfaces*—These are standards for physical interconnection of units. They are not necessarily tied to a single signal origination or bit stream standard, although some are.

4. *Signal distribution*—These standards provide a complete data format for distribution by broadcasting or cable; they include one or more video, audio, and data channels and appropriate error protection coding.

The standards discussed below fall into all of these levels.

1.6.2.1 ITU-R Rec. BT.601-5

One of the earliest digital video standards was developed by the *International Telecommunications Union* (ITU), radio communications branch, ITU-R. It is a worldwide signal origination standard that defines component digital picture formats for 525- or 625-line video systems.

The basic sampling frequency for Rec. BT.601-5 is 13.5 MHz, which is a magic number that is readily derived in either 525- or 625-line video systems. The standard provides options for sampling the three components at this frequency or, in the case of color-difference components, at submultiples of this frequency. To keep track of the sampling frequency ratios, a format of three numbers is used. Sampling at the 13.5-MHz frequency is represented by the number 4; sampling at one-half the 13.5-MHz frequency is given the number 2, and so on. An RGB system with each component sampled at 13.5 MHz is called a 4:4:4 system.

When color-difference signals are used, it is appropriate to reduce the sampling rates for the color-difference components (*subsampling*). This corresponds to reducing the color bandwidth used in analog composite systems. Thus, a YC_RC_B system with 2:1 subsampling of color-difference signals is a 4:2:2 system. Some of the other options are shown in Table 1.4. The 4:2:0 format uses vertical subsampling of the color-difference components to reduce that data by another factor of 2.

Rec. BT.601-5 formats are specified at the inputs of most units used in 525- or 625-line digital video systems. A similar approach has been used for the picture formats of the digital broadcasting standards adopted in the United States.

1.6.2.2 SMPTE 125M

The *Society of Motion Picture and Television Engineers* (SMPTE) develops standards for broadcast studios and other professional program production operations. SMPTE standard 125M [7] is a signal interface standard for parallel transmission of digital video. Being an interface standard, it defines not only the signal format, but also the cable and connectors.

The incoming data format for 125M is Rec. BT.601 digitized in the 4:2:2 format. The cable is special with 12 twisted-pair circuits and DB-25 connectors. Ten pairs are used for 10 bits/sample data, the 11th pair carries a clock signal, and the remaining pair is used for additional grounding. Cable lengths up to 50m are possible without equalization; with equalization, lengths can go up to 300m.

The 10 data lines transmit pixel data in parallel; no error protection is provided, under the assumption that it is not necessary with shielded, hard-wired connections. The color component format is YC_RC_B, according to Rec. BT.601. The components are interleaved on the 10-bit channel according to the sequence: $C_B \, Y \, C_R \, [Y] \, C_B \ldots$, where the three adjacent $C_B \, Y \, C_R$ values correspond to samples taken at the same image position (cosited) and the $[Y]$ value is the in-between Y sample that has no cosited C_R and C_B samples. This works because the C_R and C_B signals are sampled half as often as the Y signal.

Supporting the interleaving described above, the 125M clock is twice the Rec. BT.601 sampling frequency of 13.5 MHz, or 27 MHz.

Standard 125M interfacing is expensive because of the 12-pair cable, and cable length is limited. Thus, it is used primarily for interconnection of units within the same rack or group of racks. For longer distances, serial transmission is much more economical.

1.6.2.3 SMPTE 259M

SMPTE 259M [7] is a 10-bit/sample serial signal interface standard. It requires more complex terminal electronics than 125M but the single coaxial cable is much less expensive. The cable can be the standard RG-59 (or equivalent) coax already widely used for analog video connections, which is very convenient in the conversion of existing analog facilities to digital. This standard is also known as the *serial digital interface* (SDI).

Standard 259M covers both 4:2:2 components signals at 4:3 or 16:9 aspect ratios and digitized NTSC or PAL (see Section 1.6.2.5) signals. Integrated circuits are available for converting parallel signals (such as 125M) to 259M format, which hides the complexity of the serial encoding within one chip. Similar devices are available for the receiving end.

The bit rates for 259M transmission are 270 Mbps for 4:2:2-4:3, 360 Mbps for 4:2:2-16:9, 143 Mbps for digital NTSC, and 177 Mbps for digital PAL. These rates are based on 10 bits/sample; if 8-bit samples are used, they must be padded up to 10 bits by inserting two least-significant zero bits.

1.6.2.4 IEEE 1394

The SMPTE 125M and 259M are unidirectional point-to-point interfaces. A much more flexible architecture is that of a peer bus structure, where all units connect to all other units in a star, tree, or daisy chain configuration. The IEEE 1394 serial interface [8] is just that—it is a serial bus that provides two-way communication between multiple units.

A four- or six-wire shielded cable is used: two pairs of wires provide communication in each direction, and an optional third pair can supply power to units on the bus. The standard specifies the protocol for bus transmission at a number of data rates, but it does not specify anything about the data itself. Thus, the bus can be used to transmit any type or types of data as long as the data's structure is known to the units on the bus who use it.

The data rates of 1394 are specific multiples of 24.576 Mbps: ×4, ×8, and ×16. These values are close to 100, 200, and 400 Mbps and the modes are therefore called S100, S200, and S400. Data rates can be mixed on the same bus as long as the hardware supports them all. Transmission uses a packetized approach (see Section 1.6.2.6) that allows multiple users to simultaneously share the bus. A transmission mode is also available to support audio or video real-time transmission, where a specific data rate must be guaranteed so that the audio or video presentation will not be interrupted by other activities on the bus.

A number of companies have endorsed this standard and are building products that use it, especially for home and semiprofessional markets. The *1394 Trade Association* [9] has been formed to expand the use of 1394 and further its development.

1.6.2.5 Digitized Composite Video

Many digital products are designed to operate with digitized NTSC or PAL analog composite signals. This may seem paradoxical because such signals already contain the distortions of the analog composite system, which the digital system can do nothing about. However, this is advantageous in several ways.

- Once the composite signal has been digitized, it can be handled transparently by a digital system. For example, editing and multigeneration recording can be done without degradation. When the digital processing is completed, the signal can be converted back to analog and continue with the analog system.
- The digital data rate for digitized composite video is lower than the rates for component systems. Thus, handling of digitized composite video is less expensive than handling of digital component signals.

Table 1.5
Sampling Parameters for Digitized Composite Video Systems

Standard	Sampling	Bits/Sample	Data rate (Mbps)
NTSC	$3 f_{SC}$	8	85.9
NTSC	$4 f_{SC}$	8	114.5
PAL	$3 f_{SC}$	8	106.3
PAL	$4 f_{SC}$	8	141.8

- Digitizing composite video makes the unique features of digital systems, such as memory or digital special effects, available in an otherwise analog system. Some of the first digital products for television were configured in this way.

In digitized composite systems, it is generally desirable to synchronize the digital sampling rate to the color subcarrier frequency of the composite video. Sampling is usually chosen to be at 3× or 4× subcarrier frequency. Data rates for this are tabulated in Table 1.5 for NTSC and PAL.

1.6.2.6 Moving Picture Experts Group (MPEG)

Video compression is essential in many digital systems, especially home systems. It was first used commercially for storage of video on CD-ROMs and for display on personal computers. It is also a key ingredient in the new digital broadcasting standards that include HDTV (see Chapter 7). A widely used international standard was developed by the *Moving Picture Experts Group* (MPEG) subcommittee of the *International Standardizing Organization* (ISO) and the *International Electrotechnical Commission* (IEC). Currently, two MPEG standards are in use: MPEG-1 for low data rate video and audio, with limited picture quality, and MPEG-2 for full-quality video and audio from standard-definition television (SDTV) up to HDTV (at higher data rates, of course). Other MPEG versions are under development.

The MPEG standards are bit stream standards, although they also include the integration of audio, video, and auxiliary data into a single stream. Signal interfacing for MPEG may be any of the serial methods described above. MPEG bit streams are highly configurable, so there are many choices available for picture formats, number of channels in one bit stream, degree of video or audio compression, and so on. Many of these options can be changed dynamically during a program.

The compression features of MPEG are discussed in Section 7.4.4.1; only bit stream integration is covered here. Bit stream integration uses a technique called *packetizing*, a digital time-division multiplexing method that allows very flexible allocation of types of data within a bit stream. The basic idea of packetizing is to break each incoming bit stream into packets (blocks of data) and add a packet header to each one to identify it. Packets may be all the same size or they may be variable in size. In the latter case, the packet header must have a field to identify the size of each packet.

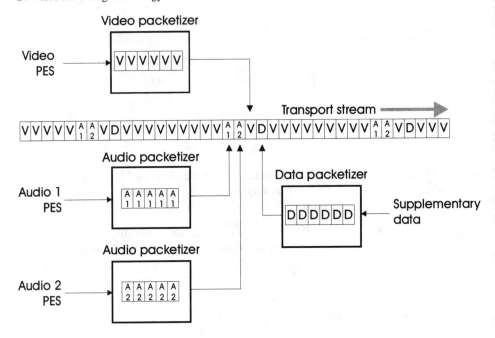

Figure 1.8 *Packetizing for transmission of multiple bit streams. Reproduced with permission from [1].*

Once one or more bit streams have been packetized, they may be integrated into a single output bit stream by interleaving them, as shown in Figure 1.8. The interleaving structure is designed to provide continuous and synchronous delivery of data in each stream. At the receiving end, a depacketizer separates the bit stream into the original streams by reading the packet headers and directing each packet to its specified destination. Some buffering is necessary at the receiver to absorb small variations of timing in the output streams caused by the packetizing.

Incoming bit streams may be combined this way as long as the total of their bit rates does not exceed the capacity of the output channel. Sometimes it is necessary to insert empty packets to pad the output signal to a specific rate. Some systems add error protection codes into the packet header.

1.7 AUDIO SIGNALS

Audio is almost always a vital ingredient in the presentation of complete video programs, and video transmission or distribution systems generally provide for one or more channels of audio to go along with the video. Multiple audio channels may support stereo reproduction, multiple languages, surround sound, and so forth [10].

Audio is much lower in bandwidth than video, since the highest frequency that is ever required is only 20,000 Hz. However, audio generally requires much higher SNR than video; audio SNR can go as high as 80 to 90 dB. The other major difference between audio and video is that audio has no structure of lines and frames as does video. Even the smallest interruption in an audio signal can be heard.

1.7.1 Analog Audio

Except for computer-generated sound, audio is picked up with one or more microphones as analog signals. Audio signals may contain multiple components of speech, music, sound effects, or background sounds. These are transmitted or processed in analog circuits that have the necessary bandwidth, SNR, and amplitude linearity. These properties are subject to analog measurement with special equipment designed for audio.

1.7.1.1 Analog Audio Performance Measures

The complexity of most audio waveforms generally precludes waveform examination of audio signals, and performance measurements are usually made by channel examination methods. The important channel properties are frequency response (bandwidth), SNR, amplitude linearity, and wow and flutter. These are tested with appropriate test signals and measuring instruments.

Bandwidth is measured using a series of equal amplitude sine wave signals covering the frequency range from 20 to 20,000 Hz. The output is measured with 0.1 dB precision or better, and the results are plotted as a frequency response curve. This may be automated by using a sweep signal generator that sweeps the frequency smoothly over the range and a measuring unit that reads the output and plots curves on paper or an oscilloscope screen.

SNR is tested on a system by measuring the output with no signal input. The noise level can be measured with a root-mean-square (rms) voltmeter that is calibrated against the normal operating level of the system. In some cases, it is useful to examine also the spectrum of the noise output with a spectrum analyzer. Because the human ear's sensitivity to noise varies significantly with frequency, a *weighting filter* is sometimes used for noise measurement to approximate the subjective effect of noise.

Amplitude linearity is tested by introducing a high-quality sine wave signal at the normal operating level of the system. At the output, the input frequency is filtered out and what remains is noise (of course) and any harmonics of the input frequency produced by distortion. Usually the noise level is low enough to be ignored compared to the products of distortion, and *total harmonic distortion* (THD) is taken as the level of the output compared to the normal signal level. Values down to 0.1% are significant. Another linearity test measures *intermodulation distortion* by inputting two frequencies and measuring the cross-modulation products produced at the output.

Wow and flutter is the result of time base instability on audio; it is frequency modulation (FM) of the audio and is measured by reading the percentage of FM on a constant-frequency audio signal. Good performance requires less than 0.1% wow and flutter.

1.7.2 Digital Audio

Digital audio systems are advantageous compared with analog systems for the same reasons described in Section 1.5 for digital video systems. Digital audio is widely used in the compact disc (CD) system for home audio. It is also used with all digital video systems and is available as an option on many analog video systems.

1.7.2.1 Audio ADC

Conversion of analog audio to digital audio must meet the same conditions described for video in Section 1.5.1. For full-bandwidth audio reproduction, sampling frequencies above 40,000 Hz must be used. A widely used value in this range is 44,100 Hz, which is used by the CD system. Using 16 bits/sample and two channels for stereo, this sampling gives a data rate of 1,411,200 bps.

Personal computer audio systems often use lower numbers for sampling rate and bits/sample, simply because of the large storage requirements for any significant amount of CD-level audio. Sampling rates as low as 8,000 Hz and 8 bits/sample have been used for speech-quality audio. This reduces the data rate to 64,000 bps. Of course, such sampling rates require a good prefilter to avoid aliasing.

1.7.2.2 Digital Audio Performance Measures

Digital channels for audio can be tested in the same ways described for video in Section 1.5.2 except that the bit rates are lower. Alternatively, a complete digital audio system can be tested by the analog methods described in Section 1.7.1.

1.8 DIGITAL TV

Most countries have or will soon choose standards for digital TV (DTV) broadcasting. This will result in the deployment of a new generation of DTV receivers that will be used not only to receive DTV broadcasts but to receive similar signals delivered by other digital means such as cable, satellite, or even high speed digital telephone networks. This section describes the DTV standards developed in the United States.

1.8.1 Background of DTV

Existing SDTV systems use standards that were developed in the 1950s and 1960s. By the early 1980s, TV researchers realized that much higher picture quality would be available in a new system based on more recent technologies. Research laboratories around the world began development along these lines and before long there were several proposals for new systems based on modern analog techniques. These systems were called *high-definition TV* (HDTV).

Demonstrations of these HDTV systems were excellent, but the systems were not considered practical for broadcasting because they required much more bandwidth than existing SDTV channels could provide. With all the demands from other services for spectrum space, it was unthinkable to ask for more bandwidth for TV broadcasting. Development continued to find ways to reduce the bandwidth needed by HDTV.

1.8.2 Grand Alliance

In the early 1990s, several proposals were made for systems based on digital technology that could broadcast HDTV in a normal SDTV channel. In the United States, the Federal Communications Commission (FCC) asked these proponents to get together and propose a single standard for digital TV broadcasting. A group called the *Grand Alliance* (GA) was formed for this purpose. Their objectives were as follows [11].

- High-quality digital HDTV pictures and sound;
- A system that could coexist with analog TV broadcasting without interference either way;
- Cost-effective equipment should be possible for consumers, producers, and all users at the time of introduction and over the life of the standard;
- Interoperability with other transmission media and applications;
- The potential for a worldwide standard.

All objectives except the last have been met by the final system design. It appears that there will probably be several versions of a digital standard around the world.

Another organization, the *Advanced Television Systems Committee* (ATSC), was charged with the task of taking the GA proposal and writing complete standards. This has been completed, and standards for the United States have been adopted by the FCC. Documents are available from the ATSC [12].

1.8.3 Layered Architecture

The DTV standards are based on the four-layer architecture as shown in Figure 1.9. The figure also compares these layers to the seven-layer structure of the ISO standard IS 7498 (1984), known as the *Open Systems Interconnection* (OSI) international standard model for digital communications protocol. Recorders are involved with the first three layers of the GA architecture. The fourth layer, transmission, is concerned with broadcasting and uses a multilevel modulation that is not applicable to recorders.

1.8.3.1 Picture Layer

This layer defines the picture scanning formats supported by the standard. Because of the objective for interworking with SDTV systems, multiple scanning formats are available. In a digital system, conversion between scanning formats is readily accomplished at moderate cost, and it is quite reasonable that the scanning format being broadcast and the

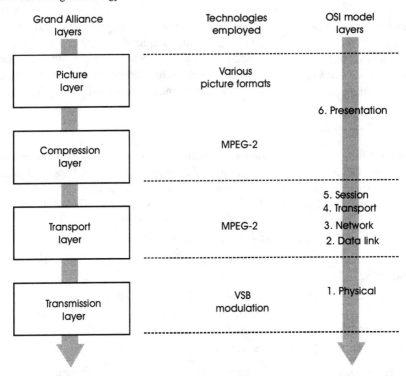

Figure 1.9 *Grand Alliance layered architecture. Reproduced with permission from [3].*

scanning format of the receiver's display may be different. Recorders for DTV, therefore, must be able to handle all of these formats.

1.8.3.2 Compression Layer

The DTV standard specifies MPEG-2 compression for all picture formats, which is an essential part of the broadcasting standards so the signals will fit the 6-MHz U.S. TV channel. Recorders may operate on uncompressed DTV signals in a production-postproduction environment to maintain the highest picture quality in these operations, and they also may be called upon to store compressed formats in less critical environments, such as broadcast station on-air operations. These may be two different recorder designs or they may be the same basic recorder with adapters.

1.8.3.3 Transport Layer

When recording compressed DTV signals, it will be most convenient to use the transport layer format, which is also taken from MPEG-2. This is a packetized serial bit stream that handles any number of video, audio, and data streams multiplexed into a single signal (see Section 1.6.2.6).

1.9 CONCLUSION

Video systems are ubiquitous in the modern world and recorders are an essential part of them. This chapter has given an overview of video system technology as it affects the design and use of recorders.

REFERENCES

[1] Luther, A. C., *Video Camera Technology*, Artech House, Norwood, MA, 1998, Ch. 3.

[2] Ibid., Section 12.2.1.

[3] Luther, A. C., *Principles of Digital Audio and Video*, Artech House, Norwood, MA, 1997, Section 2.3.3.

[4] Ibid., Section 2.4.4.

[5] Luther, A. C., *Video Camera Technology*, Artech House, Norwood, MA, 1998, Ch. 12.

[6] Benson, K. B., and Whitaker, J., *Television Engineering Handbook*, 2nd ed., McGraw-Hill, New York, 1992, Chapter 21.

[7] SMPTE standards are available from the Society of Motion Picture and Television Engineers, 595 W. Hartsdale Ave., White Plains, NY 10607-1824 or at http://www.smpte.org

[8] IEEE standards are available at http://www.stdsbbs.ieee.org

[9] The 1394 Trade Association's Web site is at http://www.1394.org

[10] Benson, K. B., *Audio Engineering Handbook*, McGraw-Hill, New York, 1988.

[11] *Grand Alliance HDTV Signal Specification*, Ver. 2.0, Grand Alliance, Dec. 1994.

[12] Advanced Television Systems Committee at http://www.atsc.org

2

Recording Fundamentals

Recording is the storage of information for later review, processing, or distribution. Video recording is the storage of pictures and sound. A familiar video recording system is motion picture film, where a series of actual pictures are stored on the film. The film can be held up to light to see the pictures or it may be "played" in a projector to present the pictures in motion.

Electronic recording stores information in forms that allow the replay of the recording to produce an electronic information signal. That signal may be displayed in its native form, processed, or distributed through copying. Pictures are converted to electronic signals by scanning, as described in Section 1.3.1. Once scanned, there is no longer any need to store the picture signal in the form of an image, as with motion picture film; many other forms become possible. Signal encoding may be either analog or digital; in the latter case, many of the techniques and equipment developed for digital computers are applicable to audio and video. This chapter discusses the fundamental methods of storing electronic signals, especially video and audio.

2.1 INTRODUCTION TO RECORDING

Electronic recording uses the physical properties of certain materials that can be permanently or reversibly altered locally to create records. The records can be replayed by detecting the alterations of the material produced during recording. The properties of magnetic or optical materials are well suited to recording, and most electronic recorders use one or both of these types of materials.

All recording methods in use today operate on the surface of the recordable material or in a very thin layer of recordable material placed on a thicker substrate that provides mechanical support. This is *area* recording—records are spread over the area of the medium. (It is theoretically possible to have a volumetric recording medium where records

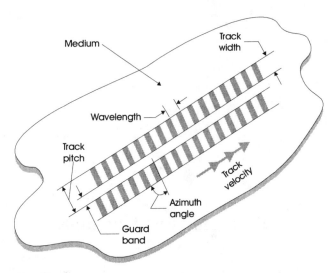

Figure 2.1 *Properties of tracks.*

would be dispersed within a three-dimensional piece of medium. So far, this has not progressed outside of research laboratories.)

An electronic signal is one dimensional—it has a single value that varies as a function of one independent variable: time. The simplest way to record this on an area medium is to make lines of records by scanning the medium. Thus, successive values of the signal are recorded along a line on the medium. These lines of records are called *tracks*.

2.1.1 Tracks

Records are produced by causing relative motion between the recordable medium and a recording device (often called a *head*). This action creates a *track* of records on the medium; the shape of the track and its dimensions are determined by the dimensions of the head and the form of the relative motion. For replay, a replay head[1] must pass precisely over each track and sense the records in the track. This is called *tracking* the records. Figure 2.1 shows the following properties of tracks.

- *Track width*—The recording head has a physical width, or a recording width, that determines the width of the recorded track. The amplitude of the replay signal is larger for wider tracks, which gives a higher SNR. The smallest practical track width is determined by the properties of the medium and heads, the SNR requirements of the system, and the mechanical considerations of scanning and tracking on replay. Tracking may be achieved by pure mechanical precision or it may involve elaborate servomechanisms.

---

1 Replay heads are not necessarily the same as recording heads. In some systems they are, but in other systems they involve different physical principles.

- *Track velocity*—The relative velocity between head and medium determines the highest frequency that can be recorded and also the rate of usage of the medium. Track velocity is sometimes called *writing speed* or *head-to-tape velocity* (for tape-based media).

- *Wavelength*—Considering the input signal as a single frequency, the spacing of cycles in the records, called the *wavelength* λ, is a function of the frequency f and the track velocity V according to:

$$\lambda = V/f \tag{2.1}$$

For example, a frequency of 1 MHz recorded at a track velocity of 10 m/s has a wavelength of 10^{-5} m. In general, recording and, especially, replay become more difficult as wavelengths get smaller.

- *Track pitch*—This is the space between corresponding points of adjacent tracks. Most applications have a multiplicity of tracks spaced together as closely as possible. However, there may be interference between tracks caused by fringing fields or tracking errors. That is avoided by leaving a little space between tracks by making the track pitch greater than the track width. The unrecorded space between tracks is called a *guard band*. (Some systems can operate without guard bands; see Section 3.2.1.1.)

- *Track azimuth*—In magnetic recording, this is the angle that the head gap makes with the track. Usually it is 90° and must be maintained with great precision. It is a consideration only when the data elements in the track are oblong or rectangular, as in magnetic recording—it is not a consideration with optical recorders having a round spot. Track azimuth error is evaluated relative to the shortest wavelength of recording; it is a $\sin(x)/x$ function of the error angle. If the azimuth positional error across the width of the track between record and replay equals the wavelength, the output will be null because of interference between signals received from different positions in the width of the track. Some systems operate with deliberate azimuth offsets to eliminate interference between adjacent tracks and thus the need for guard bands. This still requires the offset azimuth values to be held to tight tolerances between record and replay.

All systems must specify their track dimensions to assure interchangeability between recordings, recorders, and replay equipment.

2.1.2 Area Density

It is always desirable to minimize the amount of recording medium required for a given recording time because of cost and physical constraints. A system that uses less area of medium for a given amount and quality of recording is more efficient. This property is quantified by calculating the amount of recording that can be placed on a given area of medium—called the *area density* of the recording system. The amount of recording is usually specified in bits (megabits), and the area is either a square centimeter or a square

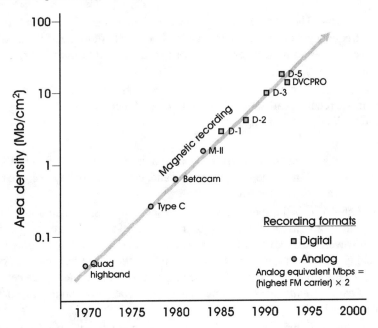

Figure 2.2 *History of magnetic recording area density.*

inch. For analog recording systems, where there are no bits, the convention is to assign 2 bits to each cycle of the highest recorded frequency when calculating area density.

Area density is calculated by the following equation:

$$\text{area density (bits/cm}^2) = 2 / \lambda P \tag{2.2}$$

where λ is the minimum recorded wavelength (cm) and P is the track pitch (cm). The factor of 2 accounts for 2 bits per cycle of wavelength.[2] For area density in bits/in^2, units of λ and P should be in inches. Area density has shown remarkable improvement during the history of recording, as shown in Figure 2.2 (for magnetic recording) and can be expected to continue improving.

2.1.3 Modulation Methods

Recording media and heads generally do not provide channel characteristics suitable for directly recording audio or video. This mismatch is accommodated by using modulation to modify the frequency and amplitude characteristics of the information signal. Analog and digital modulation methods are discussed in Sections 2.2.6, 2.2.7, and in Chapter 7.

---

2   This is an approximation for digital systems. A better value is obtained by multiplying by the density ratio (see Section 2.2.6.2).

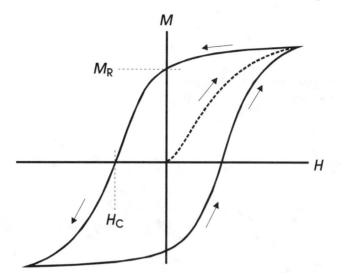

Figure 2.3 *Magnetic hysteresis loop.*

2.2 MAGNETIC RECORDING FUNDAMENTALS

Audio and video recorders use magnetic technology more than any other method. Although magnetic video recording is over 40 years old, it shows no signs of being replaced by other approaches. As of this writing, magnetic video recorders are undergoing the transition to digital, with new products being introduced at a steady pace.

2.2.1 Physical Principles

Magnetic recording is based on the property of certain materials becoming magnetized in the presence of a magnetic field and retaining some of that magnetization when the field is removed. Magnetic materials are referred to as ferromagnetic or ferrimagnetic, which signifies that they have properties similar to iron. They may be in the form of metals or of structures fabricated from powders, such as ferrites.

The property of magnetic *remanence* is shown by a curve of magnetization M versus magnetizing force H, shown in Figure 2.3. This curve exhibits a form of extreme nonlinearity known as *hysteresis*, where the behavior is different whether the magnetic force is increasing or decreasing. The dotted line shows the curve of increasing H when starting from a completely demagnetized state; the arrows show the path followed after that. If the magnetizing force is increased and decreased periodically, the curve traces a loop, called a *hysteresis loop*. Two key parameters of hysteresis loops are M_R, the remanent magnetization remaining when the magnetizing force is reduced to zero, and H_C, the *coercivity*, or the magnetic force needed to reverse the remanent magnetization to zero.

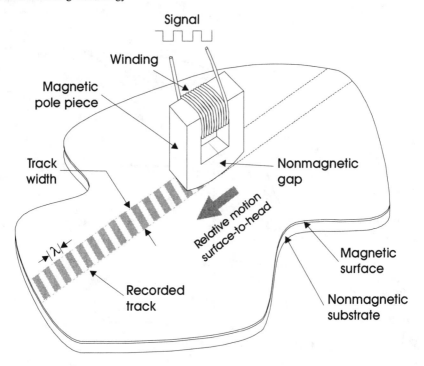

Figure 2.4 *Magnetic recording. Reproduced with permission from [3].*

The slope of the B-H curve is the *permeability* μ of the material; because of the extreme nonlinearity, it is a difficult parameter to deal with. However, the initial permeability μ_I, which is the slope of the start of the dotted curve in Figure 2.3, is more easily stated. In general, it is most convenient to deal with *relative permeability* μ_R, which is the ratio of the permeability of the material to the permeability of a nonmagnetic material. Thus, μ_R is dimensionless.

Another important characteristic of magnetic materials is the flattening of the top and bottom of the hysteresis loop. This is called *saturation*, which means that there is a maximum amount of magnetization that can be produced in the material regardless of how much magnetizing force is applied.

Magnetic recording of information uses these properties to produce a pattern of magnetization on a surface that is usually a magnetic coating on a nonmagnetic substrate. The substrate is typically in the form of a tape or disk, although magnetic surfaces are used in other forms, such as the magnetic stripe on a credit card. The combination of a magnetic surface and its substrate is called the *medium*.

2.2.2 Recording

Recording is accomplished by producing relative motion between a recording head and the magnetic medium while applying signal information to the head. Signal current in the winding of the head causes the head to create a spatial magnetic track on the surface, where the magnetization of the surface varies according to the polarity of the current in the head. This is shown in Figure 2.4. A magnetic head, consisting of a magnetic core with a nonmagnetic gap and a winding, is positioned so that it passes close to or in contact with the magnetic medium. Signal current applied to the winding of the head produces magnetization in the core, some of which fringes out at the gap into the adjacent medium. As each part of the recording medium passes through this fringing field, it is cycled through its hysteresis loop; the condition pertaining at the instant it leaves the field of the head controls the remanent magnetization that will be left at that part of the medium.

The width of the track equals the active width of the head, but the size of the pattern details along the track depends on the signal frequency and relative head-to-medium speed, as was given in the wavelength equation (2.1). Higher frequencies produce shorter wavelengths; the bandwidth limit of the system is related to the shortest wavelength that can be successfully recorded and recovered on replay.

Recording of wide bandwidth requires high-speed relative motion between the medium and the head so that the wavelength will not be too short. However, high speeds result in mechanical stresses and other difficulties. There are many ways of achieving the necessary relative speed, such as linear motion of medium past fixed heads, or rotating heads scanning a linearly moving medium, or a rotating medium scanned by moving heads. The shape of the track pattern depends on the type of motion. Mechanisms for achieving the relative speed and the motion of the medium are described in Chapters 4, 5, and 6.

2.2.3 Playback

A recorded track is read back by providing similar relative motion between the medium and a playback head. The playback head must accurately track the recorded tracks, a process that may require servo control of the motion of medium or head or both. In many cases, the same physical head is used for both recording and playback, although sometimes there are advantages to using different heads for each process. During playback, some of the fringing field from the medium is picked up in the head as the track passes under the gap region of the head. This field passes through the core of the head and a voltage is induced in the coil of the head. The voltage in the coil is proportional to the rate of change of magnetic flux in the head core, meaning that, at a given head-to-tape speed, the natural output amplitude from the head rises with signal frequency.

The gap length of a playback head acts as an aperture that averages the flux it picks up; the gap length must be shorter than the shortest wavelength in the recording. The effect of the gap length is a $\sin(x)/x$ response function, having its first null at the frequency corresponding to a wavelength equal to the gap length. In addition, there are losses due to the electrical and magnetic characteristics of the head, the separation (if any) between the medium and head, and the thickness of the recording layer.

2.2.3.1 Separation Loss

The separation effect is very significant and requires the playback head to be as close as possible to the magnetic surface. *Separation loss* in decibels is:

$$\text{separation loss (dB)} = 54.6(d/\lambda) \qquad (2.3)$$

where d is the separation between head and medium, and λ is the wavelength. For example, there is 5.46 dB of signal loss when the separation is 10% of the wavelength. For a wavelength of 3 μm, which is typical, separation must be less than 0.3 μm. It is often difficult to achieve separations that small simply because the motion of head or medium produces an air film that causes larger separation. In many cases, the head and medium have to be deliberately forced together.

2.2.3.2 Thickness Loss

Another response loss is caused by the thickness of the medium's magnetic layer. A layer that is thick compared to the shortest wavelength will direct some of the magnetization away from the surface, causing *thickness loss*. This is not as serious as separation loss, but it cannot be ignored.

The result of all these playback response effects is shown in Figure 2.5. The natural rising response from the rate-of-change of flux is overcome at high frequencies by the gap effect, separation loss, and thickness loss. The electrical response of the head, which is a resonant circuit consisting of the inductance of the coil and its stray capacitances, may further affect the total response.

The discussion so far has characterized the head in analog terms. Remembering the extreme nonlinearity of the fundamental magnetic process due to hysteresis, it can be seen that it will not be satisfactory to operate a magnetic head directly with a simple analog signal—some form of modulation must be used to produce linear signal response in spite of the inherent nonlinearity in magnetic recording. That modulation may be either analog (Section 2.2.5) or digital (Section 2.2.6).

2.2.4 Noise Sources in a Magnetic Channel

Noise sources in magnetic recording consist of medium noise, head noise, and electronics noise [2]. The importance of these sources varies with different designs, as discussed in this section.

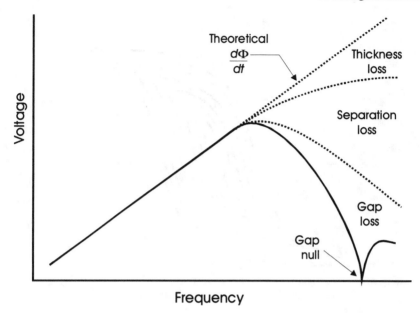

Figure 2.5 *Frequency response of magnetic recording (solid curve). Reproduced with permission from [3].*

2.2.4.1 Medium Noise

Because of the particulate nature of magnetic media, either because of actual particles in magnetic-particle media or magnetic grains or domains in thin-film media, all media generate noise. In most modern recording systems, this is the predominant source of noise, although it is trending down, and eventually, head noise may become the major source. Medium noise consists of both additive and multiplicative components.

The spectrum of additive medium noise rises at approximately 6 dB/octave with frequency because the volume of medium being sensed is directly proportional to the wavelength. This continues until the thickness loss begins to appear, then it slows down to 3 dB/octave, until separation and gap losses takes over, when both noise and signal outputs fall precipitously. These same effects cause the spectrum of maximum possible signal output to rise at a rate of 6 dB/octave at long wavelengths but to become flat when thickness loss becomes significant.

Wideband SNR is obtained by integrating the ratio of signal output power to noise power over the bandwidth of interest. It is proportional to the number of particles being sensed by the read head, more particles giving higher SNR. For a given medium design, the number of particles is proportional to the volume of a wavelength-element under the read head (gap width × $\lambda/2$ × record depth in the medium). Thus, in the absence of thickness, gap, or separation losses, SNR is proportional to wavelength.

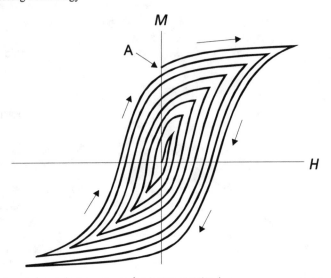

Figure 2.6 *Hysteresis loops during erasure, beginning at point A.*

Multiplicative medium noise depends on the uniformity of the medium's coating statistics along a track, including particle distribution and surface roughness (which affect separation).

2.2.4.2 Read Head Noise

Read head noise is mostly thermal noise generated according to the Nyquist noise theorem, which states: "Any device that dissipates energy when connected to a power source will generate noise power as a passive device." Thus, the real part of the head impedance is a noise source. The real part of the impedance is produced by the combination of hysteresis losses in the head core and the physical resistance of the head coil.

2.2.4.3 Electronics Noise

The read head preamplifier is the source of nearly all electronics noise. It behaves according to $1/f$, and is seldom significant in a well-designed recording system.

2.2.5 Erasing

When making a new recording, it may be necessary to erase previous recordings on the medium. An entire tape or disk may be erased by *bulk erasing*, where the medium is subjected to an alternating field that is strong enough to saturate the magnetic coating. That field is slowly reduced to zero, which cycles the medium through successively smaller hysteresis loops until both magnetizing force and magnetization go to zero, as shown in Figure 2.6.

For tape media, erasing may be done as the tape passes through a tape deck for recording by using an erase head that has a long gap length and a gap width equal to the width of the tape. This head is driven by a high frequency that is higher than would ever be recorded by the erase head. This has the same effect of cycling the medium through its hysteresis loop as it passes away from the erase head.

Selective erasing can be done on one or more tracks by using an erase head only one track in width and positioned ahead of the recording head. This method is necessary when a recording must be selectively updated in editing audio or video.

Some digital recorders do not have separate erase heads; they simply overwrite previous recordings. This is widely done with digital disk drives where it would be impractical to have more than one head operating at a time on the same medium. Some loss of recording efficiency occurs with this method; an allowance must be made for a certain percentage of signal to remain from previous recordings—this can be taken care of by digital error protection.

2.2.6 Analog Recording

Analog magnetic recorders are very successful for both audio and video. They employ either frequency modulation for video or bias recording for audio to overcome the limitations of magnetic recording technology.

2.2.6.1 Bias Recording of Audio

Analog audio recorders use a high-frequency bias signal to linearize the recording process. The theoretical explanation of this is complex [1], but an intuitive explanation can be given by thinking of the bias signal as a switching signal that rapidly switches the track area under the head from positive to negative magnetization. In the absence of any audio signal, the head acts much like an erase head, and the switching action of the bias results in zero average magnetization in the track. However, when an audio signal is present, it acts to unbalance the switching effect of the bias, with the result that the net remanent magnetization is no longer zero but is shifted in the direction of the audio signal. This might be thought of as a kind of pulse-width modulation of the bias signal by the audio signal. It is more difficult to see why this process should be linear, but it is to a high degree. Very high-quality audio recordings can be made with the proper application of bias.

2.2.6.2 Frequency Modulation for Analog Video

Most analog video recorders solve the recording channel linearity problem by using frequency modulation (FM) of the video. When correctly applied, FM provides excellent performance that is largely immune to the vagaries of the recording channel.

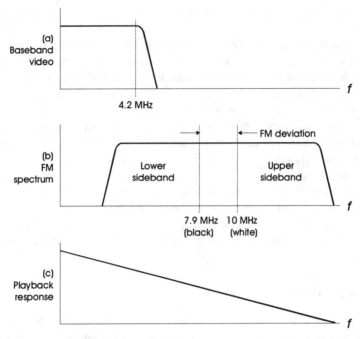

Figure 2.7 *Frequency spectrum of analog video FM.*

The hysteresis loop of a magnetic recording channel has the same characteristic as a limiter, which is a required element in an FM system. However, FM for video recording is quite different from that used in radio broadcasting in that the carrier frequencies, their deviation, and the video bandwidth are quite close together. The simplifying assumption that the frequency deviation is much less than the carrier frequency does not apply to video.

A typical frequency spectrum for video FM of an NTSC signal is shown in Figure 2.7. As shown in Figure 2.7(b), frequency modulation generates sidebands that extend either side of the frequency deviation range by an amount equal to the video bandwidth. The FM deviation is defined by the frequencies covered by the video amplitude range from black to white. (See Section 7.3.2 for more about FM.)

The system often is operated so that the recording channel response cuts off some of the upper FM sidebands, which has the effect of introducing amplitude modulation on the signal. That is removed on playback with a limiter, restoring the upper sideband at the detector. However, the sideband cutoff has to be carefully tailored to assure that the sideband energy is restored uniformly across the video bandwidth to prevent ripples in the resulting video frequency response. One approach to that is the *linear-rolloff filter*, shown in Figure 2.7(c). Such a filter rolls off the upper sideband response but has a corresponding roll-up on the lower sidebands. The result is a uniform amount of sideband energy but the distribution between lower and upper sidebands changes with frequency.

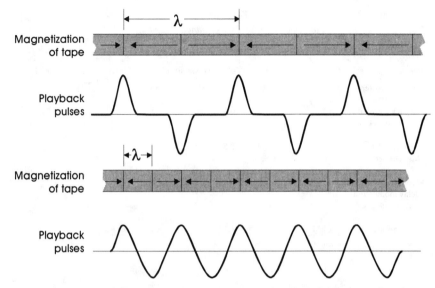

Figure 2.8 *Waveforms of digital magnetic recording showing the effect of shorter wavelengths. Reproduced with permission from [3].*

With enough channel bandwidth, NTSC or PAL signals can be passed successfully through the FM system as composite signals; this is the approach shown in Figure 2.7. However, to use less channel bandwidth, the chrominance may be separated from the composite signal and be recorded separately as an amplitude-modulated subcarrier in the low frequency range below the FM signal range. The FM signal carries the luminance and acts as a high-frequency bias for linear recording of the chrominance components. This system is used in VHS and 8-mm home video recorders; it is called *color-under* (see Section 7.3.3).

2.2.7 Digital Recording

Digital recording does not need to linearize the recording transfer characteristic when binary (two-level) signals are used. The two binary levels are simply represented by the two polarities of magnetization. Thus, bits are recorded by the transitions between flux polarities as shown in Figure 2.8. Since the playback output is the derivative of the recorded flux, the output at a flux polarity reversal is a pulse. As shown in the figure, shortening the wavelength causes the pulses to begin overlapping; once this happens, the playback output will rapidly drop for further shortening of wavelength. This effect determines the maximum bit density that can be recorded along a track. It is determined by the magnetic material and its thickness, the head properties, and the degree of frequency equalization in the playback amplifier.

Table 2.1
Parameters for Several Common Encodings

Name(s)	Acronym	T_{min}	T_{max}	DR	DC component	Self-clocking	Waveforms 1 0 0 1 1 0 1 1 1 0
Nonreturn to zero	NRZ	T	∞	1	large	no	
Frequency modulation biphase-mark code	FM	$T/2$	T	0.5	zero	yes	
Phase encoding Manchester code	PE	$T/2$	T	0.5	zero	yes	
Modified FM delay modulation Miller code	MFM	T	$2T$	1	small	yes	
Eight-to-fourteen code	EFM	$1.41T$	$5.18T$	1.41	zero	yes	(see text)

2.2.7.1 Clock Extraction

The playback digital circuits require a *clock signal* in their processing. In digital recorders, this is provided by making the data signal *self-clocking*, meaning that the clock can be extracted from the signal itself. The normal bit stream generated by an ADC is not suitable for self-clocking because there may be long strings of the same bit, meaning that the bit stream remains at the same level for a period of time. During such a period, there would be no signal for clock extraction. This can be solved by suitable encoding of the data to eliminate loss of clock energy.

2.2.7.2 Encoding

One objective of encoding (also called modulation) is to make sure that sequences of the same bit value do not become too long; a characteristic that is called the bit stream's *run length*. Run length is quantified by specifying the minimum (T_{min}) and maximum (T_{max}) times between transitions of the data stream. The T_{min} value controls the maximum frequency component in the data stream (it is approximately one-half cycle of the maximum frequency), and the T_{max} value relates to the longest time that the clock recovery circuit must hold without any input.

A further parameter is the *data density ratio* (DR), defined as the ratio of T_{min} to the minimum time T between transitions of the incoming data stream before encoding. The larger the DR, the more information is transmitted by a given channel. If T_{min} is given as a fraction or multiple of T, then DR = T_{min}.

Another important feature of a data stream is its *dc component*, which is the long term average of the bit values in the stream. This is important because most transmission media cannot transmit a dc value (this includes magnetic recorders.) Loss of the dc component of a data stream will cause errors or, at least, it will reduce the system's margin for errors. Good encoding schemes eliminate or minimize the dc component. Table 2.1 lists these properties for several common encodings.

The following list discusses each of the encodings in the table.

- *Nonreturn to zero* (NRZ) is when the ones and zeros of the data stream are transmitted directly one after another. This is the way we usually think of a raw bit stream. In NRZ, a positive-going transition indicates a one and a negative-going transition indicates a zero. Strings of repeated ones or zeros generate no transitions. NRZ is simple but impractical because the infinite T_{max} and the large dc component make clock extraction impossible.

    A variant of NRZ is NRZI (NRZ-inverted). In NRZI there is a transition (in either direction) for every one but no transition for zeros. This coding is polarity insensitive but it has the same T_{min} and T_{max} as NRZ.

- The FM encoding (also called *biphase mark coding*) transmits two channel transitions for a one and one transition for a zero. This eliminates the dc component but cuts T_{min} in half, so the DR is only 0.5. It is self-clocking.

- The phase encoding (PE) or *Manchester code* has a transition for every bit, located in the center of a bit cell. A zero has a positive transition and a one has a negative transition. When consecutive values are the same, extra opposite-direction transitions are added between bit cells. This achieves the same results as FM encoding, on dc and DR = 0.5.

- In the modified FM (MFM) code, also called *delay modulation* or *Miller code*, a one is coded by a transition of either direction at the center of a bit interval, whereas there is no transition at that position for a zero. A string of zeros will have a single transition at the end of each bit interval. This gives a DR of 1 with a probability of a small dc component.

- The *eight-to-fourteen code* (EFM) is one of many possible codes that are based on processing groups of data bits; thus, they are called *group codes* and are widely used. In EFM, the data is broken into groups of 8 bits and 6 additional bits are added to each group, so that 14 bits are transmitted for every 8 incoming bits. This insertion of extra bits is done with a lookup table that outputs 14 bits for every incoming 8-bit value. (The contents of such a table is called a *code book*.) The table has 256 14-bit entries. Because a 14-bit word can represent up to 16,384 values and the incoming data groups have only 256 values, the actual 14-bit values used can be selected to control T_{min}, T_{max}, and the dc component. Different ways to make this choice lead to different encoding systems. Of course, the method must be standardized because the decoder must know the exact code book used.

    One possibility for EFM selects 14-bit values that always have two or more like bits together. Since two output bits correspond to $(2 \times 8)/14 = 1.14$ input bits, $T_{min} = 1.14$ (T_{min} is calculated in terms of the input bits.) The DR has the same value, so by adding 6 bits to every 8 bits, we have been able to transmit 14% more data! That happens because the choice of EFM 14-bit values has avoided all single-bit transitions, thereby halving the required channel bandwidth. This EFM encoding limits the maximum sequential output bits to 7, so calculating in terms of the input bits, $T_{max} = (7{\times}8)/14 = 4$.

In practice, the EFM process is more complex than described above because there are not enough 14-bit values to meet all the conditions when considering the concatenation of adjacent 14-bit codes. To control this, the coding actually must select from a code book containing four choices for each incoming 8-bit value. The list in Table 2.1 shows the performance for the particular EFM coding used in the D-3 magnetic video recorder (see Section 10.2.1).

Encoding is further discussed in Section 7.3.4, and details for various recording products are described in Chapters 10 and 11.

2.2.8 Magnetic Media

Magnetic recording uses thin coatings or films of magnetic material supported by a non-magnetic substrate, which may be either tape or disk configurations. These are described in this section.

2.2.8.1 Magnetic Tape

A tape medium is attractive because a very large area of recording surface can be stored in a small volume by rolling tape onto a reel. On the other hand, tape is unattractive because it must be unwound to locate a particular point within a reel, which takes considerable time. Tape is most suited to applications where recordings will be played from beginning to end without stopping; it is not very applicable in cases where random access to any place in the recording is required.

With magnetic tape, a magnetic coating is placed on a flexible plastic substrate. The coating may be a suspension of magnetic particles held in a plastic binder, or it may be a metallic film deposited directly on the substrate. The magnetic properties of the coating must be maintained uniformly over the length of a tape as well as from one tape to the next. In manufacturing, uniformity is obtained by applying the coating to wide rolls of substrate, which are then slitted into numerous individual tapes of the final width. The small-area integrity of the coating must be maintained to prevent too many *dropouts* (gaps) in recording or replay. Cleanliness of the coating is also important for the same reason.

The mechanical properties of the substrate are designed to withstand the stresses of winding and unwinding as well as those of guiding the tape through the path of a tape deck, which usually depends on the edges of the tape. The elasticity of the substrate is controlled to maintain the degree of stretching under tension because tape decks have to keep the tape under tension to control its motion.

Similarly, the mechanical properties of the coating must withstand the stresses of passage over heads and other tape deck elements. The surface smoothness of the coating is critical to maintain close head contact without excessive head or tape wear, and to control friction in the tape path, which may increase if the coating is made too smooth. (A very smooth coating may actually stick to elements in the tape path.)

Many tape decks perform most of their tape guiding with elements that ride on the back of the tape to avoid possible damage or wear of the recording surface; some tapes add a special nonfriction coating on the back of the substrate to make back-side guidance more efficient. The tape must hold up both physically and magnetically for hundreds or more passes through a tape deck.

To improve the volume efficiency when tape is on reels, substrate thickness should be as small as the mechanical requirements of tape handling will allow. However, with sensitive magnetic coatings and thin substrates, there can be a problem where the recording on one layer of a reel will affect the recordings on adjacent layers—a problem called *print-through*. This is most critical with analog recording at long wavelengths. Further information about magnetic tape is given in Section 3.1.

2.2.8.2 Magnetic Disks

Disk formats are of two types: rigid and floppy. Rigid disks (also known as *hard disks*) are primarily used in digital storage devices for computers and video editing systems. A nonmagnetic metallic substrate (called a *platter* and usually made of aluminum) held to close tolerances for flatness and runout is coated with a magnetic material for the recording surface. Heads are designed to "fly" on a very thin air film so that small separation is achieved without actual head-to-disk contact (see Section 4.1.4). Noncontact operation is necessary because the disk is normally continuously rotating and the head may remain at a particular location of the disk for long periods of time; contact recording would result in rapid wear of head and disk. Because of the tiny head-to-disk separation needed for high-density recording, dust particles or other contamination at the disk surface could cause the head to "crash" onto the disk surface, which could destroy the surface or the head. Hard disks are usually sealed within a dust-free enclosure and platters cannot be changed or removed. (See also Section 3.3.2.)

Floppy disks use a flexible substrate whose coating operates in contact with the head. The compliance of the substrate and the physical positioning of the head is designed to control the forces of head-to-tape contact so there will not be too much wear. Floppy disks are normally rotated only when actual recording or playback occurs and they rotate slower than hard disks to avoid the wear of continuous long-term contact with the head. Floppy disks do not achieve the recording densities of hard disks, but they are designed to be easily removed to exchange disks between drives. (See also Section 3.3.1.)

2.2.9 Summary of Magnetic Video Recording

Magnetic video recording is a mature technology that has been in use for more than 40 years. It has been developed to suit a wide range of markets and price points, ranging from home systems costing less than U.S.$200 to professional systems costing in the tens of thousands. It will continue to be an important video recording technology for many years.

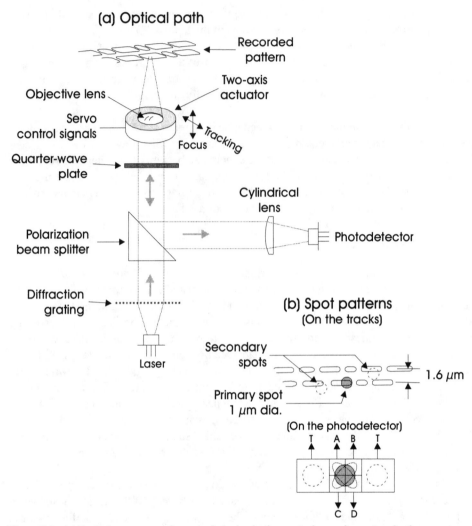

Figure 2.9 (a) *Typical components for optical playback of an optical recording using the three-spot method. (b) The spot pattern. Reproduced with permission from [3].*

2.3 OPTICAL RECORDING

Optical recording is based on media that can support patterns to be read by optical detection means. Optical recordings are not necessarily made by optical means but they are always readable by optical means. For example, the familiar audio CD is usually replicated by a mechanical pressing process but it is always played with an optical reader. There are many advantages to optical compared to magnetic technology.

- Much higher recording densities are possible; current optical systems operate at densities above 50 Mb/cm^2. The new digital versatile disc (DVD) system records at a net density of 400 Mb/cm^2 (see Section 11.2).
- Optical reading is done with a beam of light. There does not have to be any mechanical contact between the reading head and the medium because the light beam can be focused at a distance from the head. Mechanical wear is eliminated and reliable performance is possible over periods of years.
- The recorded pattern can be embedded in a plastic substrate that makes it impervious to dirt and scratches. Damage to the plastic surface is unimportant because it can be out of focus for the reading beam.
- Recordings are not affected by electric or magnetic fields.
- Read-only recordings can be made that cannot be changed after creation, which is excellent for maintaining data integrity for archival or security purposes.
- Records can be in the form of mechanical patterns that can be replicated inexpensively by a pressing process.

The disadvantages of optical recording are mainly associated with the need for more expensive recording apparatus, although the costs of these are rapidly decreasing.

2.3.1 The Physics of Optical Playback

Optical playback is based on the properties of monochromatic light as generated by lasers. A laser beam can be focused to a very fine spot and it can be made to interfere with itself as a way to sense very small changes in physical dimensions or optical transmission and reflectance. Regardless of the type of recording method, all optical recordings are read in much the same way, as shown in Figure 2.9(a).

A solid-state laser generates a beam of monochromatic light that is focused by an objective lens onto the track pattern, which usually consists of a series of spots where the normal optical properties of the medium have been altered by recording. Several components of the optical path are involved in a tracking servo, which will be ignored in this first explanation. A polarization beam splitter in the optical path passes the laser light to a quarter-wave plate that rotates the plane of polarization by 45°. At the track pattern, the light is reflected back toward the source, returning through the same optical path. A second pass through the quarter-wave plate adds an additional 45° rotation of polarization, with the result that the polarization light splitter reflects the return beam over to a photodetector instead of passing it back to the laser. As a result of the changed optical properties of the medium in the recorded spots compared to unrecorded areas of the medium, the light beam is reflected at a different amplitude when it strikes a recorded spot than when it strikes a space between recorded spots. Thus, it becomes amplitude-modulated by the record pattern. This modulation of the return beam to the photodetector is then detected and processed to recover the information signal.

2.3.1.1 Optical Tracking and Focus Servo

The dimensions of the recorded pattern are extremely small—submicron in some cases. Servomechanisms are required for the laser beam to focus and track the recorded pattern to compensate for tolerances in the manufacture and mounting of the disk. The servo method described here is known as the *three-spot method*, and can be understood with the help of Figure 2.9(b).

A diffraction grating placed in the optical path near the laser creates an interference pattern where the laser beam is split into a series of repeated spots of decreasing amplitude. This is shown in Figure 2.9(b). Only the main spot and the first two side spots are used for detection of both focus and tracking. The objective lens is mounted in a two-axis electromechanical actuator that allows it to be moved in a direction along the optical path for focusing and a direction perpendicular to the optical path for tracking.

Focus is detected with the main spot, which changes in shape as focus shifts as a result of the cylindrical lens being in the path of the photodetector. This lens introduces *astigmatism* distortion, which makes the spot become elongated when it is out of focus. The photodetector that receives the return spots is actually divided into three segments as shown in the figure; the optical system is set up so that the main spot falls on the center segment and the two side spots fall on the other two segments. The center segment is further divided into quadrants. When the system is in focus, the main spot will fall equally on each quadrant. However, when it is out of focus, the astigmatic spot will energize the quadrants unequally as shown by the dotted lines in the figure. By appropriately processing differences between the four output signals from the quadrants, an error signal is developed for control of focus. The modulation on the main spot is detected by summing all signals from the four quadrants.

Tracking is detected by the two secondary spots falling on the side segments of the photodetector. As shown in the figure, the side spots are aligned so they fall off each side of the main track. As tracking shifts, those spots will be amplitude modulated by the recorded patterns of adjacent tracks. When the degree of modulation of the side spots is equal, it will be known that the main spot is on-track. The amplitude unbalance between the side spots controls the transverse actuator of the objective lens to maintain tracking.

2.3.1.2 Modulation Methods for Optical Recording

The optical recording process just described can be used with either analog or digital modulation, although all new optical products are using digital modulation.

Analog modulation of the optical recording process is achieved by modulating the length of the spot patterns along the track. This can be thought of as a form of frequency modulation, which can be detected by measuring the length of the patterns to recover the information signal. Early optical recording systems, such as the laser videodisc, use this approach.

Digital modulation can be applied to the recorded pattern by several methods, including a spot–no spot approach, spot length modulation, and so on. Since the optical detection is basically digital to begin with, it is an excellent digital recording system.

2.3.2 Optical Recording Methods

Early optical recorders used a high-power laser to actually burn holes in a metallic coating. This gave rise to the terminology of *pits* to describe the recorded areas and *lands* for the unrecorded spaces between the pits, because the pits were actual holes. For manufacturing CDs by pressing, a negative master is made by plating over the recorded pattern of pits and lands and then stripping off the plating, with the result that the pits became "bumps." The negative master is used in a pressing machine to replicate copies in hot thermoplastic material. The pressing process is still used to replicate mass quantities of the same discs at extremely low cost per disc.

However, development continued to make the original recording process less expensive and practical for one-at-a-time recording. Recorders are now available for a few hundred dollars as a PC peripheral. These use blank discs that have a pressed pattern on them for tracking during recording. The actual recording is made with lower laser power, at a level that can be obtained by a low-cost solid-state laser. The coating material contains a dye material that changes color or optical phase shift under the recording beam; such recordings are read by the same playing method as just described (see Section 3.3.3).

This technology has now been further extended to create media that are *rerecordable*, meaning that recordings can be erased and done again at the same position on the medium. These are called *rewritable* optical devices.

2.3.3 Area Density of Optical Records

Data density along a track is similar between magnetic and optical recording (in the range of 4,000 and 20,000 bits/cm). However, track pitches in optical recording are 10 to 20 times smaller than in magnetic recording (400 tracks/cm for magnetic and 4,000 to 8,000 tracks/cm for optical). This largely results from the optical recorder's round recording spot. Magnetic recorders have a rectangular recording "spot" resulting from the dimensions of the head gap, which are limited by mechanical considerations in head and tracking system design. The result is that optical area densities are 10 or more times higher than magnetic recorders. The density curves are trending upward for both technologies as magnetic heads and materials improve and laser wavelengths are shortened, but it is unlikely that magnetic density will catch up with optical density any time soon.

2.3.4 Optical Media

All optical video recording systems to date use a disk format for the medium. Disks have a number of advantages over tapes.

- A disk is a simple one-piece mechanical structure that does not require any reels or other mechanical components for its use. Some disk formats enclose the medium in a housing or carrier to protect it from handling, but this is still simpler and inexpensive compared to a tape cassette that serves the same purpose for tape.

- The entire recordable surface of a disk is immediately accessible for record or replay. Thus, random access to any location in a recording is fast and easy.
- The mechanism (disk drive) for recording and playing a disk is much simpler than that of a tape deck.

With these advantages, all recorders might use the disk format except that the surface area of a disk is inherently limited by $\pi(R_O^2 - R_I^2)$, where R_O and R_I are the outer and inner radii of the recordable area of the disk. With tape media, the recording area can be made larger simply by increasing the width or the length of the tape. In both formats, there are practical limits of dimensions determined by media package size and recorder size, but tape media can provide much greater recordable area within these limits.

In earlier days, video recording densities were such that a disk could not provide a reasonable amount of recording time (usually considered to be at least 1 hour per each media package), so tape was the only approach that would work. Today, with optical area densities reaching more than 400 Mb/cm^2, a single disk medium, with digital video compression, is capable of recording an hour or more of video.

2.3.4.1 The Compact Disc

The first optical recording system to reach mass marketing was the Compact Disc[3] for digital audio (CD-DA), which uses a 12-cm diameter disc recorded on one side; it is capable of 74 minutes of digital recording of high-quality uncompressed stereo audio. The first application, prerecorded audio, used a mastering process to produce discs replicated by pressing. Hundreds of millions of CD-DA players have been manufactured and they are now priced well below US$100.

The use of the CD as a general digital storage device with a data capacity per disc of up to 680 MB was immediately recognized, and standards for a computer version were developed under the name *CD-ROM*. This also has seen widespread use and virtually all PCs manufactured today either have a built-in CD-ROM drive or it is available as an option. CD-ROMs were originally produced by the same mastering process as the CD-DA; but recently, CD-recordable (CD-R) drives have become available for PCs at very attractive prices. These are capable of recording CDs one at a time using the capabilities of an ordinary PC.

A still more recent development is the CD-rewritable (CD-RW) drive, using a CD medium that can be recorded, selectively erased, and rerecorded. More details of all the CD formats can be found in Section 11.1.

---

3   The normal spelling of this word is *disk*. The CD marketers chose to call their product *Compact Disc*, which is a trade name. This spelling is used when the word refers to the CD; in all other cases, it is spelled *disk*.

2.3.4.2 DVD

Since the original development of the CD around 1982, optical recording technology has significantly improved in area density capability. Thus, it was soon realized that a more advanced format would eventually be practical. Development toward this goal was started. The combination of video performance and playing time of the present CD-ROM format is marginal, even with the best compression available. For 1 hour video recording time, the picture quality is marginal, and for good picture quality, the playing time is too short. Operating at a higher density, it would be possible to deliver extremely high-quality video with playing times per disk of 3 or 4 hours. Developers in several companies have reached this goal, and there has been extensive standards activity to bring the developments together in a single standard. This has been completed as of this writing and products are now available for initial marketing.

The product is called the *Digital Versatile Disc* (DVD). The acronym also sometimes is interpreted to mean *digital video disc*. Marketing of DVD has begun for both computers and the home. DVD will surely become widespread and will probably replace the CD-ROM and other formats over the next 5 to 10 years. More details of DVD are available in Section 11.2.

2.4 CONCLUSION

Video recording technologies, both magnetic and optical, have shown vast improvement over the years since their introductions. This is based on growth in the capabilities of supporting technologies such as integrated circuits and digital techniques, as well as continuing development in recording systems, materials, and components.

REFERENCES

[1] Mee, C. D., and Daniel, E. D., *Magnetic Recording, Volume I: Technology*, McGraw-Hill, New York, 1987.

[2] Ibid., Section 5.2.

[3] Luther, A. C., *Principles of Digital Audio and Video*, Artech House, Inc., Norwood, MA, 1997.

3

Tapes, Disks, and Records

All recording standards define the patterns that are placed on the medium in the act of recording, as well as the modulation format of those patterns. The presumption is that any equipment capable of recording the specified patterns is standard, regardless of how it is physically accomplished. Similarly, any equipment that can play the specified patterns is also considered a standard player. The patterns are called records; this chapter discusses the types of media in detail and covers the ingredients of records and the considerations of their use. Specific record standards are described in Chapters 10 and 11.

3.1 TAPES

A tape medium makes available the largest area of surface for recording. When recording densities were low, tape was the only way to achieve a long recording time for audio or video. However, with current recording densities, especially with optical recording, the disk medium has come into use for recording audio and video. The future is in disk formats. Note that a tape medium has not been used for optical recording; tapes are only magnetic.

3.1.1 Tape Construction

A magnetic tape consists of a substrate and one or more coatings. Figure 3.1 shows a cross-section of a typical magnetic tape. A plastic substrate is made of *polyethylene terephthalate* (PET, also known as Mylar, a tradename of DuPont), with thickness in the range of 10 to 40 µm. The substrate provides the mechanical strength of the tape and a smooth base for the coatings.

The back coating is usually a carbon-based material that controls the coefficient of friction for the back surface that may run on tape guides in a tape deck. It also reduces the resistivity of the tape, minimizing the tendency for static electricity to build up during fast reeling.

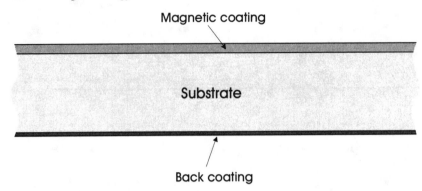

Figure 3.1 *Cross-section of a typical magnetic tape.*

3.1.2 Magnetic Coatings

The magnetic coating is composed of either magnetic particles suspended in a nonmagnetic binder material or a thin film of metal. The coating thickness depends on the type of use—analog audio tapes that require longer wavelength recordings usually have thicker coatings than video or digital tapes, where the modulation methods eliminate the need to record long wavelengths.

3.1.2.1 Particulate Coatings

Various materials have been used as magnetic particles. The most widely used material until recently is *gamma ferric oxide* (γFe_2O_3). These particles are *acicular*, about 0.2 to 0.3 µm long with an aspect ratio of about 6:1. The magnetic properties of the tape are optimized by *orienting* the particles to have their long dimension in the direction that tracks will take on the tape. That is accomplished during manufacture by applying a suitable magnetic field to the tape while the coating is being applied but before being cured.

The binder material, when cured, holds the particles in place, provides a wear surface on the tape, and may contain a lubricant to smooth the head-to-tape interface. An important consideration of binders is to prevent binder components from collecting on heads or the tape deck and interfering with the proper head-to-tape interface. Sometimes, small amounts of abrasive material are added to the binder system to help keep the heads clean. Of course, that must be carefully controlled to avoid excessive head wear.

Other particulate materials include *chromium dioxide* (CrO_2), pure iron (Fe), or *barium ferrite* (BaFeO). These provide higher coercivity than γFe_2O_3 and are used in some of the latest high-density video and digital tapes. High coercivity has its problems, however, because it may be difficult to utilize the high coercivity without encountering saturation of the head pole tips during recording (see Section 4.1.1.1).

Tape design and manufacture is a highly developed art [1].

3.1.2.2 Thin-Film Coatings

Thin magnetic films are often used for magnetic hard disks; they are less common on flexible-substrate media such as tape and floppy disks. This is because of the difficulty of achieving a layer that will withstand the bending of the substrate inherent in a flexible medium. There may also be problems of head wear or corrosion of the metal film surface.

Thin films may be manufactured by electroplating, radio-frequency (RF) sputtering, or evaporation. The materials include cobalt, chromium, amd alloys of these and other metals. A thin film tape or disk often includes several layers of different materials to build up the necessary magnetic and mechanical properties (see Section 3.3).

3.2 MAGNETIC TAPE RECORDS

This section covers the considerations of magnetic tape records. Many of these factors also apply to magnetic disks, covered in Section 3.2.2.

3.2.1 Magnetic Tracks

A group of magnetic tracks is shown in Figure 3.2 (this is a repeat of Figure 2.1, reproduced here for convenience.). These may be either *longitudinal tracks* (parallel to the edges of the tape), or *helical tracks* (placed at an angle to the edge of the tape). A third type, *transverse*, is a special case of helical tracks at an angle approximately 90° to the tape edges. These were used in the first videotape recorders, known as *quadruplex* recorders, which were used for many years in broadcast and professional applications but now are no longer being manufactured. Today, essentially all videotape recorders use helical tracks.

The general properties of tracks for all types of recording were discussed in Section 2.1.1; this section enlarges on that discussion for magnetic tracks.

3.2.1.1 Track Width

All magnetic video records are in the form of tracks. The most important parameter of a track is its width, which is determined by the design of the recording and replay heads. Thus, track width is a characteristic that cannot easily be changed once a system is manufactured because it is determined by mechanical dimensions of the heads. Some systems, especially longitudinal recorders, provide plug-in heads that can be exchanged to modify track width, but this is unusual.

Since wider tracks increase the area of the medium used for a given piece of recording, the signal output from a replay head is increased in proportion, everything else being equal. Increased track width will always increase channel SNR; the degree depends on what is the most important noise source in the channel (see Section 2.2.4).

Factors other than SNR may limit the minimum track width. In replay, it is necessary for the replay head to accurately follow the recorded track to avoid loss of signal

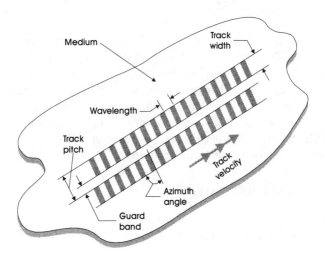

Figure 3.2 *Magnetic recording tracks.*

or interference from adjacent tracks. In most longitudinal recording tracks, replay tracking is controlled by purely mechanical tolerances of head positioning and tape guiding. Tracking errors are in the range of 20 to 100 μm, depending on the precision of tape edges, tape width, and positioning of heads and tape guides.

Replay tracking of helical tracks is controlled by servomechanisms (see Section 5.2) that precisely set helical scanning speed and tape speed. These servos generally make use of a longitudinal *control track* at the edge of the tape. The success of this approach still depends on many mechanical dimensions of tape deck and helical scanner, and manual adjustment is needed to take up those tolerances. The limit of track width with this type of control is in the range of 100 μm. To provide operation of the scanner at other than normal tape speeds for slow motion or high-speed playback, and because of other problems in interchanging helical tapes, such as track straightness, many helical recorders have an additional servomechanism that watches head output and adjusts the lateral head position dynamically for the greatest output. With this approach, track widths of 20 μm or below can be used. Note that some recent digital helical recorders have even eliminated the need for the control track and tape speed servomechanism.

Because there will always be some amount of tracking error, means must be provided to prevent interference from adjacent tracks for a reasonable amount of tracking error. With longitudinal tracks or early designs of helical tracks, a *guard band* is provided between tracks by making the pitch of the tracks greater than the track width. This often amounts to as much as 20% of the track width, which is a deliberate loss of area density in favor of tracking reliability.

More recent helical-scan recorders usually are designed for *azimuth recording*, where the azimuth angles of adjacent tracks are offset from 90° to prevent adjacent-track interference. This works as long as the track modulation is designed to prevent the recording

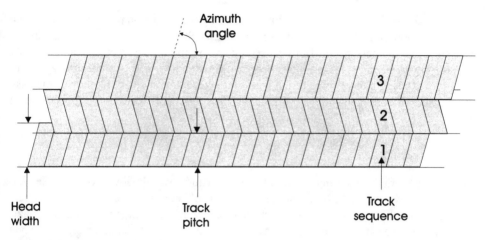

Figure 3.3 *Azimuth recording of helical tracks. Reproduced with permission from [6].*

of long wavelengths; a feature that is easily achieved in either analog FM or digital video recorders.

Azimuth recording is shown in Figure 3.3. It works on helical recorders that have an even number of recording heads on the scanner drum. The head widths are greater than the desired track width so that adjacent tracks overlap slightly during recording. The head azimuth of each of the pairs of heads is offset in opposite directions. A typical amount is ±7°. As each head passes across the tape, it records over a little of the previous track, so there is no guard band. On replay, the heads may be the same width or they may be narrower; in either case, when a head reads over an adjacent track from its own, the resultant 14° azimuth error provides nearly complete cancellation of the adjacent track's data.

In some recent designs of digital home helical recorders, the heads are made nearly twice as wide as the resultant tracks. The azimuth cancellation effect is good enough that each head picks up only its own track, regardless of where it tracks the recorded pattern, so no tracking servomechanism is needed. However, a tape speed servomechanism may still be needed for proper control during editing.

3.2.1.2 In-Track Density

The minimum recorded wavelength determines the in-track density. There are limits to short wavelengths in both the medium and the head design. As the wavelength is reduced, a smaller area of medium is involved in recording each cycle, which reduces the signal output. At the same time, the head gap length must also be reduced to keep the head operation away from the gap null. The result is that SNR reduces faster with wavelength shortening than it does with reducing track width.

Head efficiency, manufacturability, and reliability are reduced by shorter gap lengths, so this ultimately becomes a limit too.

In-track density performance is limited also by the head-to-tape separation effects. Even with supposedly "in-contact" recording, there will be residual separation because of tape surface roughness and air films caused by relative motion between tape and head. At very short wavelengths, these effects become important. Air film separation may be reduced by exerting greater force of the head on the tape, but this very likely will increase the wearing of both tape and head. In the end, it is a tradeoff between performance and head or tape life.

3.2.1.3 Track Azimuth

The use of azimuth offsetting to eliminate guard bands was already covered in Section 3.2.1.1. However, regardless of whether azimuth is offset or not, there is a critical tolerance on azimuth between recorded tracks and replay heads. This exists because an azimuth error between track and replay head causes a phase shift of picked-up signals as one moves incrementally across the head width. Such phase shift will cause partial signal cancellation in the total output of the head. The amount of phase shift depends on how the accumulated dimensional azimuth error (not the angle) across the track width relates to the wavelength of the track pattern. The dimensional error d is calculated from the angle θ and the track width w:

$$d = w \tan(\theta) \tag{3.1}$$

The behavior of azimuth error is according to a $\sin(x)/x$ function having its first null at $d = \lambda$. When d is one-quarter λ, the azimuth loss is approximately 1 dB. However, when exchanging tapes between machines, it should be noted that azimuth errors can occur separately in record and replay, so the effect can be doubled.

3.2.2 Tape Speed Issues

The recording time of a tape system depends on the total length of tape available and the tape speed. If the tape length is L and the tape speed is S, the maximum recording time T_{MAX} is:

$$T_{MAX} = L / S \tag{3.2}$$

Longer tape length for a given tape thickness implies larger tape reels. Reel size ultimately becomes limited by equipment size, weight, or cost, and by the maximum number of layers of tape that can be piled on a reel because of accumulated tension effects (the inner layers of a reel of tape can experience excessive stresses under environmental changes).

In longitudinal recorders, the tape speed and the track speed are identical. Thus, the maximum bandwidth performance is directly proportional to the tape speed. Practical tape speeds for longitudinal recorders are determined by tape handling considerations and reel size limits (because of the requirement for a certain amount of recording time per tape). These factors limit practical longitudinal tape speeds in the vicinity of 2.5 m/s. That number as a track speed is far too low for any kind of video recording. All videotape recorders use helical tracks for the video channel because the video track speed can be made much larger than the tape speed (see Section 3.2.4).

3.2.2.1 Tape Reel Capacity

The approximate tape length capacity of a given tape reel can be calculated from the inner radius R_I, the outer radius R_O, and the tape thickness t (all in the same units):

$$L = \pi (R_I + R_O)(R_O - R_I) / t \tag{3.3}$$

This equation assumes that there is no air trapped between layers of tape on the reel, which is usually valid if the tape is reeled at the correct tension. Reel capacity can be increased by increasing the outer radius, reducing the radius of reel hubs, or reducing tape thickness. Each of these has practical limits.

- *Outer radius* is limited by considerations of equipment size, weight, and cost. For example, there is a practical limit on reel size if a tape deck is required to fit in a 19-in. rack cabinet.
- *Reel hub radius* causes excessive tape stress if it is too small. The tape can become permanently deformed when stored on too-small reel hubs for a long time.
- *Tape thickness* is limited because thinner tape has less mechanical strength and is more difficult to handle in a tape deck.

Of course, the same considerations mentioned here apply to tape stored in cassettes. In addition, the size and weight of a tape cassette is affected by the quantity of tape it holds.

3.2.3 Tape Stretch Issues

Magnetic tape, being an elastic medium, is going to stretch. It will change dimensions with temperature, humidity, and tension; these changes affect track pattern dimensions and shapes and contribute to tape interchangeability problems. Although high quality tape media is designed to minimize these effects, they cannot be totally eliminated, so tape deck and system design must take them into account.

Numbers for deformation of different kinds of tape are not given here. There is more discussion of this subject in Section 5.2, and further information and a bibliography can be found in [2].

3.2.4 Helical Tracks

Helical scan mechanisms are described in Section 4.2. This section discusses some of the properties of helical tracks that apply to all types of helical mechanisms.

3.2.4.1 Helix Angle

The governing parameter of a helical track is the helix angle α, which is the angle that the helical tracks make with the tape edges when the tape is stretched out on a flat surface.

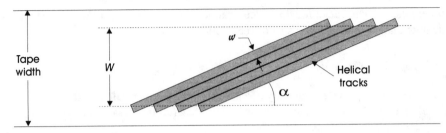

Figure 3.4 *Helical track dimensions.*

This is shown in Figure 3.4. If the helical track width is w, measured perpendicular to the track, and the helical track rate is R per second, the tape speed S is:

$$S = wR/\sin(\alpha) \tag{3.4}$$

This assumes there are no guard bands. The track speed V of the helical track depends on α, and the actual tape width used by the helical tracks. This latter factor occurs because helical tracks usually do not extend all the way across the tape width; space is left at the tape edges for longitudinal tracks. Thus, if the effective tape width of the helical tracks is W, then:

$$V = WR/\sin(\alpha) \tag{3.5}$$

and the ratio of the helical track speed V to the tape speed S is:

$$V/S = W/w \tag{3.6}$$

which is, somewhat surprisingly, independent of α. The significance of α is that smaller angles create longer helical tracks, which also require a larger-diameter scanner drum.

In analog recorders, the helical tracks are designed to hold one field of the video signal to avoid the possible distortion of switching heads during the picture, so the track rate R is 59.94 Hz for NTSC or 50 Hz for PAL. The helix angle must be chosen to provide a long enough track for 1/59.94 or 1/50 sec of recording.

In digital recorders, there is no reason to have one track per field and heads can be switched anywhere without concern for distortion; shorter tracks are often used simply because that requires a smaller helical scanner mechanism. However, for editing reasons, it is still desirable for the helical scanning to be synchronized with the video field rate. When there are multiple tracks per field, it is called *segmentation*.

Tape is an elastic medium and, at the tensions used in tape decks, there is significant stretching of the tape. Thus, the actual helical track length on a relaxed tape is slightly less than the value obtained from considering the scanner diameter and angle of operation.

3.2.4.2 *Track Straightness*

When a tape is laid out smoothly on a flat surface, all standards call for helical tracks to be straight lines. However, when the tape is in the tape deck and under tension, the tracks may no longer be straight if there is nonuniform tension across the width of the tape. This can happen because of misalignment of the tape path or because of nonuniform friction as the tape passes around a helical scanner. Achieving straight helical tracks is an important problem in helical recorder design.

Some helical recorders have a servomechanism that can dynamically move the head in a direction perpendicular to the track while the helical scanner is in motion. This feature is used to allow playback at other than normal speed, for special effects such as fast or slow motion (see Section 5.5.5). However, this feature also can compensate for track straightness errors.

3.2.4.3 Helical and Longitudinal Tracks

Most helical recorders have some longitudinal tracks along the edges of the tape for control, audio channels, or cue data or message channels. It is mechanically difficult to prevent the helical scanner from crossing over these tracks during recording, and special means must be taken to protect the longitudinal tracks. This may take the form of gating circuits that turn off the video record current when the heads pass over the longitudinal track positions, or the longitudinal track heads may be located beyond the helical scanner in the tape path so they are recorded after the video, thus overwriting the helical tracks. With the latter approach for high-quality analog audio on longitudinal tracks, the audio track space must be erased before recording to present any residual effect of the video tracks on the audio track.

In editing, it is often necessary to record video or audio separately. The strategy for protecting the audio tracks from the video tracks must stand up in such audio-only or video-only editing situations.

In digital recorders, audio and video are easily placed together in the helical tracks. This is usually done by assigning a portion of the helical track length to audio as shown in Figure 3.5, which is the track pattern of the SMPTE D-3 digital video recorder. Four audio channels are provided, with the data for two audio tracks at each end of the helical tracks. Three longitudinal tracks provide for servo control, time code, and cue audio. Thus, the audio can be edited separately by turning on the record head only as it passes the predetermined location(s) for blocks of audio data. Video-only editing uses the same approach but turns on the record heads at the track positions assigned to video. With this approach, there is no need to use longitudinal tracks for program audio, and the digital multiplexing can support more audio tracks for such purposes as multiple languages, and so on.

3.2.5 Track Measurements

In recorder manufacture, and occasionally when mechanical parts have to be changed during maintenance, it is necessary to measure track positions to verify that standard recordings are being made. A quick check is possible by playing a *test tape*, which is a tape known to be made close to standard, but a better check is to mechanically measure the recorded track positions under a microscope. This is possible by developing the

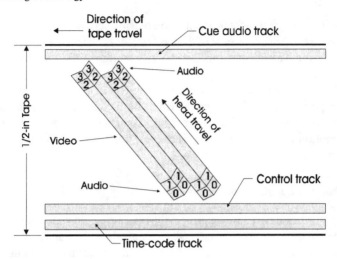

Figure 3.5 *A typical digital recorder track layout using the helical tracks for both audio and video (SMPTE D-3).*

magnetic patterns on the tape with a fluid containing a dispersion of magnetic particles that will align themselves with the track patterns when the fluid is spread on a tape. A visible image of the tracks is created, which can then be measured under a microscope that has a precisely calibrated moving stage. Both longitudinal and helical track positions can be verified this way.

3.3 DISKS

Disk media consist of a single recording surface that rotates under a head capable of moving radially to access most of the disk's surface. Tracks are either spiral or concentric. Magnetic disks have either a flexible substrate (floppy disk) or a rigid substrate (rigid disk, hard disk, fixed disk). Optical disks have either a semi-rigid or rigid substrate. This section discusses the construction of disks and their track configurations.

3.3.1 Magnetic Floppy Disks

To a first approximation, one could cut circles out of a wide magnetic tape to make floppy disks. However, several factors cause magnetic floppy disks to be quite different from that. Floppy disks generally use a somewhat heavier substrate (up to 0.003 in) although the same PET material is used. The heavier construction is necessary for stability when the disk is rotating. Disk substrates also must have isotropic mechanical properties so that tracks will remain round under mechanical or environmental stresses.

Identical magnetic coatings are usually applied to both sides of a floppy disk, and floppy disk drives have a separate head for each side of the disk, which doubles the storage capacity compared to a single-sided disk. The two heads are held by the same positioning arm, but they are slightly offset from each other so the compliance of the disk provides a controlled head-to-disk contact force. This in-contact operation could cause rapid wear of disk and heads except that floppy disk drives generally rotate the disk only when data is to be written or read.

The most important difference between tape and disk manufacture is that disks must provide uniform mechanical and magnetic properties as the disk rotates. For example, the longitudinal orientation generally applied to particulate tape coatings would cause a severe rotational variation if such tape were cut into disks. Coatings for floppy disks must achieve their performance without relying on orientation.

3.3.2 Magnetic Rigid Disks

Most rigid disk substrates are made from aluminum, although small amounts of other materials are usually added to improve the mechanical properties. Magnesium, manganese, and chromium offer improved structural strength and corrosion resistance. Substrate thicknesses range from 0.2 to 1.2 mm for the small rigid disk assemblies made for PC use.

The substrate must be fabricated to achieve maximum flatness and a good surface finish. Even so, the substrate finish is not good enough for directly supporting the magnetic coatings; an electroplated layer of nickel up to 0.1 mm thickness is generally applied to smooth out the variations in the aluminum surface. This is important for the close head-to-disk separations that are used. Although rigid disk heads are not technically "in contact" with the disk surface, they are separated by an air film that may be as small as 0.05 μm; even a minute distortion of the disk surface could cause a head crash or excessive wear. As opposed to floppy disk drives, rigid disk drives generally are rotating continuously, ever ready for rapid access to data.

The extremely smooth disk finishes necessary for "flying" heads (see Section 4.1.5.1) can cause a stiction problem under power-off conditions when the disk stops rotating and the heads settle onto the disk surface. Heads can stick so tightly to the smooth surface that they will prevent restarting of the disk rotation. To avoid this, most disks have a special area where the surface has been deliberately textured. The disk drive controls are designed so that the heads are always moved to this area before stopping or starting of the disk rotation.

Modern rigid disks use a *sputtered* magnetic layer composed of several metals. In sputtering, an ion beam in a vacuum is directed at a target fabricated from the coating ingredients. Molecules of the target are dislodged and move toward the substrate under an applied electric field. Landing on the substrate, a film grows whose composition is identical to the target material. The process can be controlled to provide very accurate and uniform depositions.

Several other layers complete the disk coating, as shown in Figure 3.6. The nickel undercoat has already been mentioned, but it is usually desirable to place a thin layer of chromium over the nickel to create a good surface for adherence of the sputtered

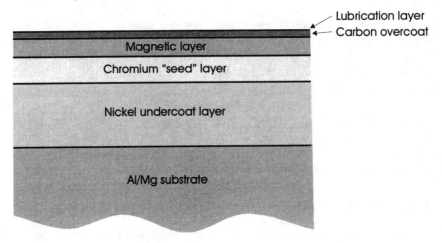

Figure 3.6 *Cross-section of a thin-film rigid magnetic disk showing layers.*

magnetic layer. This is called a "seed" layer. The wear resistance of the magnetic layer is greatly improved by adding a thin layer of a hard material such as carbon (diamond) over it. Finally, a very thin lubricating layer can further improve the mechanical performance of disk and heads. These protective layers are thin enough that they do not affect the disk's magnetic performance.

3.3.3 Optical Disks

The fabrication of optical recording disks is quite different from that of magnetic disks [3].

3.3.3.1 Optical Substrates

Optical substrates are either glass or plastics, with plastics rapidly taking over the market. Plastic substrates are pressed in a mold from the liquid state; they come from the molding machine with surfaces good enough for direct application of the sensitive coatings. Note also that the Compact Disc replication process replicates substrates and recorded patterns in the same operation. That explains the low cost of CD-audio or CD-ROM copies.

The optical transmission properties of disk substrates are important because the light beam always passes through the substrate. Light transmission should be high; in reading, the light beam passes twice through the substrate. With polycarbonate material, which is most commonly used with CD-sized disks, transmissions of 90% or better are achieved.

Another important optical property of the substrate is *birefringence* (also known as *double refraction*). This is the property of certain crystalline materials of having different indices of refraction for differently polarized light. When this effect is present in the sub-strate, it interferes with the proper focusing of the light beam. Thus, it must be kept low in substrates for optical recording. Birefringence increases with substrate thickness, so it is one factor arguing for thinner substrates. The offsetting factors, however, are mechanical—strength, flatness, and rigidity.

3.3.3.2 Optical Coatings

In optical replay, a laser beam is focused on the recording and reflected back from a mirror layer. The return beam is separated from the incident beam in a polarizing light splitter (see Section 2.3.1) and is detected by a photodetector. As a result of the recorded properties of the disk, the return beam is modulated by the information signals. Different disk systems use different coatings.

- *CD and CD-ROM*—These are produced by pressed replication from a master disc using a plastic molding process. The record patterns are a series of pits and lands pressed into the surface of the substrate. These discs have a deposited mirror layer over the pressed patterns and a thick protective layer over that, as shown in Figure 3.7(a). The patterns are read through the substrate; the disc label, if any, is placed over the protective layer on the back.

- *CD-R (recordable CD)*—These have a translucent layer of an organic dye that becomes opaque when heated by the recording laser beam. The substrate is pressed initially with a spiral groove that is used to position the laser beam during recording. A back coating of gold provides the high reflectivity needed for this process. Figure 3.7(b) shows a cross-section of a CD-R disc. As with other CD discs, an overcoat layer protects the recording surfaces and provides for a disc label.

- *CD-RW (rewritable CD)*—Construction of these discs is similar to CD-R discs but the dye layer is replaced by a material that changes its state from high reflectance to low reflectance or back again with application of the appropriate laser power. Specifically, a low laser power only reads the layer, a much higher laser power will make the medium go to low reflectance, and an intermediate power serves to return the medium to its normal reflective condition [4].

- *Magneto-optical (M-O)*—This is a system that combines optical and magnetic techniques. The recordable layer is a magnetic material having a low *Curie temperature*, which is the temperature above which the material loses its magnetic properties. During recording, a laser beam locally heats the material above its Curie temperature; existing recordings are erased but a separately applied magnetic field will be recorded as the material cools. This is a reversible process, so the material can be rerecorded over and over. Since the recording is magnetic, recordings must be read using M-O sensing means. One such method uses the *polar-Kerr effect*, where the polarization of a light beam reflected from an M-O layer changes slightly depending on the magnetization in the layer. An optical system that senses this slight change in polarization retrieves the signal.

Other coatings are used in optical recorders of both the write-once or write-many forms. However, none of these is significant in the marketplace compared to CD-based formats.

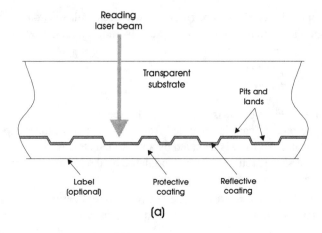

(a)

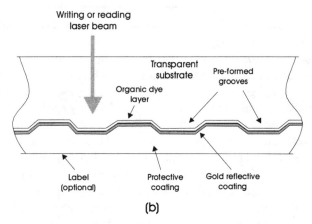

(b)

Figure 3.7 *(a) Cross-section of a pressed CD-ROM disc (along a track) and (b) cross-section of a CD-R disc (perpendicular to tracks).*

3.3.4 DVD Discs

Although the basic recording methods are the same, the construction of the new *Digital Versatile Disc* (DVD) is significantly different from the CD-based discs described above. Whereas CD discs have only a single recording layer, DVD discs support up to four layers in the same disc. This is achieved by using both sides of the disc for recording and allowing either one or two layers on each side. The two layers are accessed by changing the focusing of the read laser, which also means that the reflective coating of the first layer (closest to the surface of the disc) must be only semireflecting so that some of the light can pass through it when the second layer is being addressed. Fabrication of this multilayer structure is accomplished by using a separate substrate for each side of the disc and bonding them together after depositing the appropriate layers to produce the finished disc.

3.4 DISK RECORDS

Disk records are based on only two track configurations—spiral or concentric. Spiral tracks are used in systems basically designed for uninterrupted recording over many rotations of the disk, such as is the case for audio and video. The CD format, which was originally developed for prerecorded digital audio, uses spiral tracks.

Concentric tracks are used in systems designed for random access to small segments of information; this is the format of all computer hard disks. Hard disks can still record or play continuous audio or video by the use of buffering to cover up the short interruptions that occur when the recording or playing head moves from track to track. Random access operation is also possible with spiral tracks, but it may be a little slower than with concentric tracks. Thus, the CD-ROM, which has spiral tracks, operates effectively as a mass storage device that performs just like a slightly slow hard disk.

The analog laser videodisc system has both concentric track and spiral track modes. In the concentric mode, the disc rotation is synchronized with the TV signal at frame rate; track changes occur during the vertical blanking interval (VBI). The advantage of concentric tracks here is that the head motion can simply be stopped to produce a still picture by reading the same track (frame) over and over. Also, the disc can be used as a still picture storage unit by recording each track separately with a different still picture. This is useful in many image database applications of the videodisc system.

3.4.1 Rotation and Track Speed

It is obvious that the track speed on a rotating disk varies with the radius of the track. At the same signal frequency, outer tracks record longer wavelengths than inner tracks.

3.4.1.1 Constant Angular Velocity

A system that uses a constant rotational speed (*constant angular velocity* [CAV]) must be designed to meet its performace goals on the innermost track. Outer tracks will have excess performance that often cannot be used. Computer hard disks and the concentric track mode of the laser videodisc are CAV systems.

3.4.1.2 Constant Linear Velocity

The wasting of performance capability in CAV has led to some systems that change their disk rotational speed to maintain constant track speed as the track radius changes. These are *constant linear velocity* (CLV) systems and they can achieve higher data capacity per disk than CAV systems. For example, the laser videodisc capacity doubles in CLV mode.

CLV requires a servomechanism for the disk drive motor, capable of changing rotational speed as the track radius changes. This is a complication and it may introduce a time delay when using a CLV system for random access. The problem has been overcome in modern CD-ROM drives by running the disk at a high, constant rotational speed and making the read head work at any track speed. Digital buffering is used when necessary to change the apparent output data rate. (Most digital systems are happy to receive data as fast as possible and can handle data rate variations as part of their internal processing.) All CD- and DVD-based systems are CLV.

3.4.1.3 Sectors

A single disk track may hold a lot of data. In the case of a spiral, one track can hold the full capacity of the disk; for concentric tracks, each track holds a fraction of the disk capacity depending on the total number of tracks. To support ready access to any location on the disk, even within one track, most disks are divided into *sectors*. These are blocks of uniform data size that can be identified by a physical location on the disk.

In the case of concentric tracks with CAV, there is a fixed number of sectors per track, and thus, a given sector within a track is associated with an angular position. The physical address of a particular piece of information would be given by its track number and sector number. When that information is recorded, its physical address is stored in a *directory*, which is a table (on the disk) of information names and disk physical addresses. If a particular piece of information is contained in several tracks, the directory would have all the information necessary to find it.

A spiral track is also divided into sectors, but the physical position on the disk of a particular sector in this track must be calculated by processing an algorithm for the disk layout. If the disk is CAV, sectors can still be a specific fraction of a disk rotation, as with concentric tracks. The algorithm for that would simply be to divide the sector number by the number of sectors per track to find the track number. However, if the disk is CLV, the calculation is more complicated because the number of sectors per track varies with the track's radial position. Either way, the calculation of physical address is not difficult for a computer.

3.4.1.4 Access Time

A major advantage of disks compared to tape is that random access to any part of the recording is easy and fast. However, "fast" is seldom fast enough in a digital system, and it is important to understand the limits of random access for disks, usually stated by an *access time* specification. There are a number of ingredients.

- *Pre-seek analysis*—This is the process of translating a request for data into the physical address on the disk. It is done in the source device (computer), which, it is assumed, holds a table somewhere in memory for this purpose. If that is true, the computation time is probably negligible. However, sometimes the table of addresses is on the disk itself, in the form of a directory. If the directory is needed, a complete access cycle must first occur to retrieve that.
- *Head seek*—Having the physical address in hand, the first step of accessing is to move the head or heads to the correct track. Commands to the head positioning

servo are issued and a certain amount of time (the *seek time*) will be necessary for the head to reach the desired position. This time will depend on how far the head must be moved from its current location to the new location. This could be just a single track, or it could involve a sweep across the entire disk surface. Figures for seek time are usually given as averages.

- *Track acquisition*—Once the head is in position, data is read from the track under the head. The head positioning mechanism of magnetic hard drives is usually designed to reliably deliver the desired track in the seek process, but that is not the case in optical drives. In that case, the tracking and focus servos must lock on the nearest track, which takes some time. Even after that, the track may not be the desired one because the mechanical accuracy of the head positioning alone is not good enough with the high track density of optical disks. Therefore, the data must be read from the disk to check the sector numbers that are being read. Each sector has header information added to the data for this purpose. When the sector IDs are in the correct range, the system can just wait for the exact sector to come along as the disk rotates. However, if the sector number is beyond the desired one, the system must issue a reverse seek of one or more tracks and again check the sector numbers. This can add substantially to access time and it again is a statistical quantity.

- *Rotational latency*—Once the correct track has been found, it is just a matter of waiting for the disk to rotate to the angular position of the desired sector to begin reading data. This is known as *rotational latency* and it also is statistical, depending on just where the disk rotation is when track acquisition has been completed.

- *Data decode*—Finally, the data is read from the disk, but the modulation format must be decoded to perform error correction and put the data into the format for use. This may involve complex processes such as decompression, and significant delays are sometimes encountered. However, this also is part of the access time.

Several of the above items are statistical in that the time of access depends on where the heads were at the beginning of access and what the angular position of the disk was at the time the heads landed on the desired track. Thus, any statement of access time must also be statistical. Usually, average values are given; rarely is a statement given for maximum access time. Large data objects may span many sectors; if these are contiguous on the disk, the access simply runs from one sector to the next. However, often it is impossible to have contiguous sectors because of bad blocks on the disk or other problems; in such cases, data retrieval is slowed by the necessity for multiple seeks.

3.4.2 Magnetic Disk Tracks

Tracks on magnetic disks face all the issues discussed in Section 2.1.1 for longitudinal and helical tracks. However, there are some further considerations peculiar to magnetic disks that are covered here.

3.4.2.1 Tracking Errors

Significant tracking errors can result from the rotation of the disk and the positioning of heads. Magnetic disk drives generally do not have an actual track-reading servomechanism such as described in Section 3.2.4.2 for helical tracks or in Section 2.3.1.1 for optical tracks. They depend heavily on mechanical accuracy to achieve tracking. Most modern hard disk drives contain several platters for data recording, all running on the same spindle. A single head positioning mechanism positions a head on each platter surface. Most drives devote one platter surface to a servomechanism based on reading special reference tracks prerecorded on that platter (see Section 6.2.2.3). This can compensate for some of the mechanical errors, but not all. The possible sources of error in disk head tracking include the following:

- *Disk runout* is an off-center rotation of the disk that may be caused by either the centering of the disk on the spindle or the spindle itself. The result is that tracks vary slightly in radius as they rotate, so a head fixed in position will not read the track uniformly throughout disk rotation. A servomechanism reading a reference set of tracks can compensate for spindle runout, but drive performance is still subject to the differences between runout behavior of the various other platters.

- *Temperature* causes dimensional changes in disks and head mechanisms. Some of these can be compensated for in the design, but residual variations may cause tracking errors. Again, the head positioning servo reduces these problems down to the level of differences between platters and heads.

- *Head positioning errors* result from the inability of the head positioning mechanism to precisely repeat its position every time a particular track is requested. This should be small in a well-designed drive, but there will still be some residual errors caused by small differences in the mechanical stability of individual parts.

The accumulation of all these errors between recording and replay should not be allowed to add up to more than about 20% of the track width nor should it cause the replay head to cross over into adjacent tracks. Thus, the minimum track width and guard band in a disk drive are limited by these mechanical considerations. Drive performance has improved over the years as design and manufacturing approaches have improved; today's highest-performing magnetic disk drives operate reliably with track densities approaching 4,000/cm.

3.4.2.2 Track Speed

Track speed is determined by disk rotation rate and track radius. The trend is to smaller disk radii and higher rotational speed. This arises because of competition on the basis of drive storage capacity; to a lesser extent, access time; drive size; and cost. Faster rotational speed increases recorded data rates and reduces access time. Smaller disks can be rotated faster, but they have less recording surface. There are many trade-offs.

Table 3.1
Optical Recorder Track Dimensions

	CD	DVD
Track pitch (μm)	1.6	0.74
Minimum pit length (μm)	0.83	0.4
Laser wavelength (nm)	780	650
Disc diameter (mm)	120	120
Substrate thickness (mm)	1.2	0.6

3.4.3 Optical Disk Tracks

Optical disk densities are limited by the wavelength of the laser light.

3.4.3.1 CD Tracks

Low-cost systems like CD-based products use an infrared laser and achieve 6,000 tracks/cm. The in-track density is somewhat higher because the guard band between tracks is not involved in the calculation; it is about 12,000 bits/cm. Multiplying these numbers gives an area density of almost 72 Mb/cm^2. CD-based track patterns are shown in Figure 2.9 and dimensions are in Table 3.1.

As already explained in Section 2.3.1, these track dimensions require servomechanisms for both focusing and tracking. Hardware for this has been highly developed and CD-ROM drives are now available at retail for under $50. With all this development has come the potential for even higher densities, which is the basis for the new DVD system.

3.4.3.2 DVD Tracks

DVD track dimensions are also shown in Table 3.1. By increasing both the track density and the in-track density, the data capacity of the same-sized disc is increased by a factor of seven for a single-layer disc. With all four layers in use, the capacity of a DVD disc is 15.9 GB—24 times that of a CD-ROM.

The track density for DVD is 12,000 tracks/cm and the in-track density is 25,000 bits/cm. This gives an area density of 300 Mb/cm^2, which is about four times the CD density. The rest of the 7× improvement of data capacity in DVD is obtained through better signal processing techniques.

Notice that the thickness of each DVD substrate is one-half that of a CD. This allows two substrates to be bonded together to produce a final DVD disc having the same thickness as a CD disc. However, this causes a problem in designing drives to play both CD and DVD discs because the DVD records are 2:1 closer to the surface than CD records. Such drives must have two different optical systems to play both formats [5].

3.5 CONCLUSION

This chapter has reviewed the construction of magnetic tape and the construction of disks, both magnetic and optical. The considerations of the record patterns (tracks) used on both tapes and disks were also covered.

REFERENCES

[1] Jorgensen, Finn, *The Complete Handbook of Magnetic Recording,* 4th Ed., McGraw-Hill, New York, 1996, Ch. 11–14.

[2] Ibid., pp. 331–335.

[3] Williams, E. W., *The CD-ROM and Optical Disc Recording Systems*, Oxford University Press, Oxford, 1996.

[4] Purcell, L., and Martin, D., *The Complete Recordable-CD Guide*, Sybex, San Francisco, 1997.

[5] Shinoda, M., et al., "Optical Pickup for DVD," *IEEE Transactions on Consumer Electronics*, Vol. 42, No. 3, August 1996, pp. 808–813.

[6] Luther, A. C., *Principles of Digital Audio and Video*, Artech House, Inc., Norwood, MA, 1997.

4

Heads and Helical Scanners

In a recorder, a *head* is the device that transduces between the recorded patterns on the medium and electrical signals. In a magnetic recorder, the head is usually a simple composite device, whereas in an optical recorder the head consists of a complex electro-optical assembly embodying a laser, lenses, servomechanism actuators, and various other components. This chapter discusses both types of heads.

A helical scanner is a special type of head assembly that records and plays helical tracks on magnetic tape. The second section of this chapter covers scanners.

4.1 MAGNETIC HEADS

Figure 2.4 showed a rudimentary magnetic head consisting of a magnetic core with a nonmagnetic gap that picks up fringing fields from the tape, and a coil that senses the flux passing through the core. This is the basic concept of most inductive-pickup magnetic heads, but details of specific designs vary widely in their attempts to meet a number of conflicting goals.

- *Small dimensions*—As tracks have become narrower and wavelengths shorter, the corresponding head dimensions have followed suit. Some heads are smaller than the head of a pin. Manufacturing of such small devices is a real challenge.

- *Close tolerances*—The small dimensions of heads must be held to extremely close tolerances to meet the requirements of performance and interchangeability between recorders and tapes or disks. That adds to the manufacturing challenge of small dimensions.

- *Manufacturability*—Modern magnetic heads go far beyond the capabilities of human fabrication; manufacturing processes must be automated to deal with such small, precise parts. To meet cost goals, many processes are designed to batch produce heads or head parts; the technologies of semiconductor manufacturing have been put to good use in head manufacture.

- *Long life of head and tape*—Many heads operate in close contact with tape moving at high relative speeds. The potential for wear of both head and tape is enormous; selection of materials and the factors governing head-to-tape contact forces must be closely controlled in both head and mechanism design. Heads must operate for thousands of hours in this service. Head and mounting design must also provide for the inevitable replacement of worn-out heads.

- *High signal performance*—Above all the other aspects of head design and manufacture is the need to achieve and maintain a signal performance that is close to the highest theoretically possible. This generally drives the dimensions to be smaller and more precise and the material specifications more demanding.

These and other factors make head design a highly specialized art that combines many engineering and manufacturing disciplines.

4.1.1 Types of Magnetic Heads

For the purpose of this discussion, magnetic heads may be considered in three broad ranges:

1. Inductive-pickup heads that respond to the rate of change of magnetization in the plane of the record medium; called *longitudinal recording*;
2. *Magnetoresistive heads* that respond to magnetization in the plane of the medium;
3. Inductive-pickup heads that respond to the rate of change of magnetization perpendicular to the plane of the medium; called *perpendicular or vertical recording*.

The inductive-pickup and magnetoresistive structures for longitudinal recording are shown in Figure 4.1. Perpendicular recording, which has not made much impact in video recording, is not covered further. (See [1].)

4.1.1.1 Inductive-Pickup Heads

The purpose of an inductive-pickup head is to intercept some of the fringing field from the record pattern and direct it through a coil that produces an electrical output. A nonmagnetic gap in the pole tip is necessary to bring the flux lines into the core. As Figure 4.1(a) shows, the core picks up most of the fringing flux from the tape, but a little of it simply crosses the gap and does not flow into the coil. This occurs to the extent that the magnetic *reluctance* of the gap (reluctance is the magnetic equivalent of electrical resistance) is high compared to the reluctance of the path through the core. This loss is unavoidable. In recording, an inductive head operates in reverse—recording currents in the coil produce a fringing flux at the gap, which magnetizes magnetic coating of the record medium.

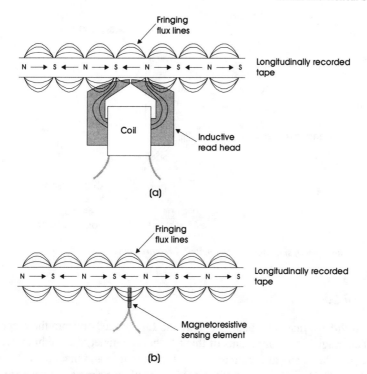

Figure 4.1 *(a) Inductive and (b) magnetoresistive pickup with longitudinal magnetic recording.*

4.1.1.2 Magnetoresistive-Pickup Heads

A magnetoresistive (MR) head senses the actual flux, not the rate-of-change of flux as with inductive heads. A thin piece of MR material is placed near the record surface so that flux lines pass through it. Its resistivity changes by a few percent as the flux changes; this is sensed by passing a small dc current through the material and picking up the voltage with an amplifier. Certain alloys of iron and nickel (FeNi) display magnetoresistivity. Because the effect responds only to the magnitude of flux, not the polarity, it is necessary to magnetically bias the MR element to obtain response to changes of flux polarity. This may be done with a permanent magnet or by the magnetic field from a coil carrying dc. The MR effect cannot be used for recording. Systems using MR generally have separate inductive heads for recording. Various designs that combine both types of head in a single assembly have been developed (see Section 4.1.5.3).

4.1.2 Materials for Magnetic Heads

Figure 4.2 is a drawing of a rudimentary inductive magnetic head showing the material requirements for different parts of the structure. Each of these is discussed below.

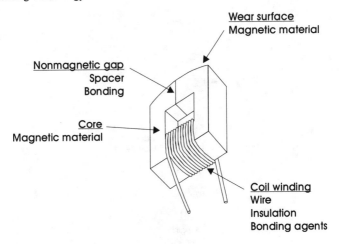

Figure 4.2 *Magnetic head, showing different material requirements.*

4.1.2.1 Head Core Material

The core provides a magnetic path to couple the flux picked up from the recording with the coil. The magnetic requirements of this are that the material have high permeability and low losses. Mechanical requirements include strength and machinability to support the manufacture of head parts and, if the core material also serves as the surface contacting the record medium, wear properties suitable for this (see Section 4.1.2.2). Table 4.1 lists electrical and mechanical properties of a number of magnetic materials often used in heads. The properties in the table and how they affect head design are discussed in this list.

- *Permeability*—This should be as high as possible in the frequency range of interest. For video applications, which always involve high frequencies, ferrite materials are best. The objective for any core design is that the reluctance for the magnetic path around the core should be low compared to the reluctance of the parallel path through the nonmagnetic gap of the pole tips. As with parallel electrical resistors, the parallel reluctances of the paths through the gap and the core determine the division of flux shown in Figure 4.1(a). Magnetic reluctance is:

$$R_M = X / \mu A \tag{4.1}$$

  where X is the length of the magnetic path, μ is the permeability, and A is the average cross-sectional area of the magnetic path. Thus, core path length should be minimized by making the core only large enough to accommodate the winding. Also, permeability and core area should be high. In a simple design, the core thickness is determined by the track width, but many narrow-track heads have wider cores that are narrowed only at the pole tips, to gain a greater core area for most of the path length. This reduces reluctance and improves mechanical strength.

Table 4.1
Electrical and Mechanical Properties of Magnetic Materials

| | Metals | | Ferrites | | |
Property	Permalloy	Sendust	MnZn	NiZn	MnZn sgl cryst.
Permeability					
1 kHz	20,000	30,000	12,000	850	10,000
1 MHz	40	60	2,000		1,500
10 MHz			700	500	700
Coercivity (Oe)	0.05	0.05	0.04	0.4	0.03
Saturation (G)	9,000	11,000	5,000	4,000	5,000
Resistivity ($\mu\Omega$-cm)	25	90	10^7	10^{10}	10^{10}
Curie Temp. °C	460	500	150	125	170
Vickers hardness	130	530	650	600	800
Thermal (μm/m/°C)	13	15	11	9	9

- *Coercivity*—This should be low in a head core material to prevent magnetization of the core, which could cause noise in playback or possible damage to records.
- *Saturation*—This is important only for record heads or heads that will be used for both record and replay. Saturation must be avoided during recording, and the higher the coercivity of the magnetic coating, the higher the flux density required in the head core. The effect of saturation is that the head is unable to fully magnetize the magnetic coating, resulting in a weak recording that has poor signal performance. The low saturation flux values of ferrites can be a problem when ferrite heads are used with high-coercivity record media. Because saturation occurs first in the gap region, some ferrite head designs overcome the problem by using a high saturation metal in the gap region.
- *Resistivity*—This affects the amount of eddy current losses in the core material. Higher resistivity is better. The ferrite materials excel in this respect; metal materials usually have to be cut into thin laminations to overcome resistance losses.
- *Curie temperature*—This is the temperature above which the material loses its magnetic properties. It is not a problem with metals but can be limiting with ferrites if cooling of the heads is not considered. Fortunately, at high head-to-medium speeds, where there could be a lot of heating, there is usually sufficient moving air to provide the necessary cooling.
- *Vickers hardness*—The hardness of the material is one of the factors affecting head wear. Higher hardness is better, although if it is accompanied with brittleness, there can be problems with breakage or eroding at high head-to-medium speeds.

- *Thermal expansion coefficient*—Thermal expansion is important in that all the parts of the head should be closely matched for expansion coefficient to prevent mechanical failure of the head structure at temperature extremes.

These descriptions show that there are many trade-offs in selecting head core materials.

4.1.2.2 Wear Surface

The surface of the head containing the gap must be in close proximity to the recording surface or in contact with it. In the latter case, wear of both head and recording surface is an important consideration. Considering the ratio of reluctances of gap and core, the *depth* of the gap (perpendicular to the record surface) is made as small as possible consistent with mechanical strength. This reduces the cross-sectional area of the gap in the direction of the magnetic path and raises its reluctance. Often the gap depth is as small as 25 µm.

Wearing on the head surface reduces the gap depth; if that is too small or the wear rate is too high, the head will have a short life. Failure under these conditions occurs when the gap depth has been worn away, causing the gap length to rapidly increase. That causes catastrophic loss of high frequency response during replay.

Three other considerations of pole tip wear are important: erosion, smearing, and gap proud. Erosion occurs when the gap spacer material is softer than the core material so that the gap wears down faster than the pole tip. This can cause core material at the edges of the gap to break away, which can be a serious problem with ferrite cores. The effect is to widen the gap. Other core materials, especially soft metals, may not break away when unsupported by the gap spacer; instead, they begin smearing across the gap, narrowing it.

An opposite effect occurs when the gap spacer material wears slower than the core material. Then, the gap becomes proud of the head surface, lifting the record medium away from the head and causing increased separation loss. In an extreme case, gap spacer material can break off, which may cause severe damage to the recording surface.

4.1.2.3 Nonmagnetic Gap

Some of the considerations of gap spacers were given in the previous section. Because the gap spacer needs no magnetic characteristics, its requirements are primarily mechanical. It is seen that the wear characteristics of spacers must match that of the core material in which they are embedded. For metal cores, gap spacers can be nonmagnetic metals or alloys. One that has been popular is *beryllium-copper*. Most spacers today, however, are materials that can be deposited on the pole tips by evaporation or sputtering, such as silicon monoxide (SiO), silicon dioxide (SiO_2), or aluminum oxide (Al_2O_3). These processes give the necessary control of the small thicknesses required. Some head designs apply spacer material to both pole tips and then use a thermal process to bond the two pieces together using the spacer material as glue, but most heads use *glass bonding* where fillets of glass are added below and at the sides of the gap area.

4.1.2.4 Coil Winding

Except for thin-film heads (see Section 4.1.5.2), many head coils are physically wound through the hole in the completed pole tips. This is an extremely tedious operation, generally done under a microscope and requiring much skill. When done by hand, this is one of the most expensive parts of head assembly. Thus, there has been a lot of emphasis on the development of thin-film heads, where the coil is produced by semiconductor manufacturing processes such as etching or deposition. Even so, thin-film heads have not been very successful for tape heads, although they are now the preferred technology for magnetic hard disks where the heads "fly" over the disk surface (see Section 4.1.5.1).

Coil wire can use any of the high-conducting metals or alloys, copper being the most widely used. The wires are very thin, and insulating materials must be equally thin. Special wires have been developed for use in magnetic heads.

The remainder of Section 4.1 covers specific head applications and the different considerations that apply to them.

4.1.3 Longitudinal Tape Heads

Longitudinal tape heads operate at low head-to-tape speeds and are used in audio and instrumentation recorders, and for the longitudinal tracks of helical video recorders. Most of these applications involve direct recording for such things as control tracks or digital tracks for time code, or bias recording for audio or instrumentation tracks. These applications generally call for longer wavelengths than FM or digital video tracks. Heads for long wavelengths must have high permeability at low frequencies and should have pole tips longer that the longest wavelength in the direction of tape travel. The latter is to prevent the leading and trailing edges of the head from acting as gaps and recording or picking up long wavelengths. Laminated metal pole tips are generally used.

Figure 4.3 shows typical construction of a laminated-core head structure having two or more tracks. The heads are mounted in a comb structure that provides support to the tape passing over the heads. As the figure shows, this is a machined nonmagnetic carrier that holds the heads and provides a curved surface to interface with the tape. A lower-cost design uses an epoxy compound to fill the space around the heads; that is finished to a tape wear surface. This is not as good as a metal surface because epoxy has a tendency to pick up and collect foreign material from the tape.

Head-to-tape contact on longitudinal tape is achieved by "wrapping" the tape around the head by an angle of a few degrees. The tape guidance in the tape deck provides the wrap angle, and the force generated by the tape tension keeps the tape in contact with the head. The radius of the head's tape-supporting surface must be a little less than the wrap curvature to assure that the tape forces will bear at the head gap region.

Multitrack longitudinal heads require shielding between the heads, and it is usually not possible to build heads as close together as the tracks on a multitrack tape can be. As a result, multitrack recorders often have two head stacks with each stack holding alternate tracks. That gives room between the heads for good shielding. It is also good practice to

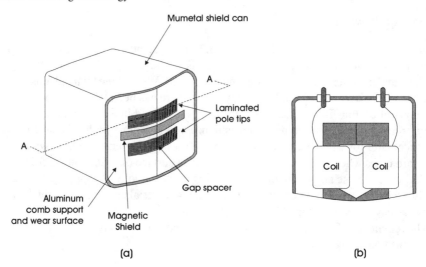

Figure 4.3 *Laminated-core longitudinal tape head construction for two parallel tracks: (a) external view and (b) section A-A.*

enclose the entire head stack in a shielding can made of high permeability material such as mumetal. Some recorders even extend the head shield to the other side of the tape; this may require some kind of moving door design.

An important consideration in multitrack tape systems is maintaining the alignment of all the heads to be on a line perpendicular to the tape edges. This is something like azimuth alignment, but it is called *skew* alignment and is usually guaranteed by building the entire head stack in halves that are each lapped to a flat surface before joining or clamping them to produce the final assembly. (*Lapping* is an abrasive machining process that is widely used in fabricating head parts.) Most multitrack tape decks will require adjustment features to be built into the head stack mountings so that heads may be set up for precise alignment to the tape edges.

4.1.4 Helical Scan Heads

Heads for helical scan systems have quite different requirements from the typical longitudinal-track heads: they operate at much higher head-to-tape speeds; they usually must handle shorter wavelengths and they do not need to handle long wavelengths; they must withstand the mechanical shock of entering and leaving the tape on each rotation of the scanner drum; they must be built for azimuth recording; and they need to be physically small and low in mass to fit on the head drum. Many developments over the years have solved all these problems; the current best designs use single-crystal ferrite pole tips with deposited gap spacers and glass-bonding to hold the head parts together. Figure 4.4 shows a simplified diagram of the manufacture of such heads.

Pole tips are made from bars of ferrite material that are machined to produce a number of heads from each bar. Two bars are bonded together to make a series of heads side by

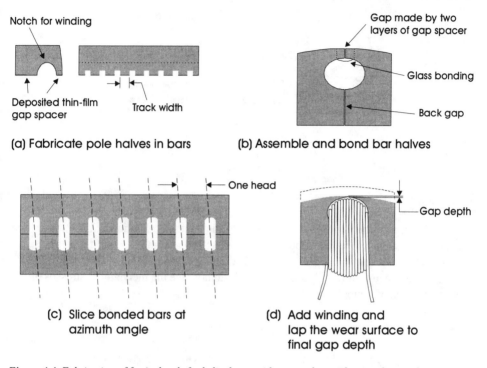

Figure 4.4 *Fabrication of ferrite heads for helical-scan video recorders with azimuth recording.*

side. This assembly is then sliced at the azimuth angle to produce individual heads that have their gaps at the proper offset angle. As shown, the head cores are wider than the final track width, with the actual width produced by a notch in the vicinity of the gap. This notch is filled with bonding glass so that the tape will not be sucked into the notches, which could damage the tape and interfere with good contact at the gap. After individual heads have been cut from the bars, coils are added and the gap area is lapped down to the required gap depth.

Helical scan heads usually are mounted to a carrier support that allows the head to be fastened to the drum with a screw that provides for adjustment or future replacement when heads wear out. The head itself is usually glued to the carrier, and various features of the carrier and drum provide for head positioning axially and radially on the drum (see Section 4.2.2.1).

4.1.5 Rigid Disk Heads

As already mentioned, the heads on a rigid disk do not usually contact the disk surface during use. They are designed to "fly" on a thin air film a specific distance from the recording surface of the disk. The air film is a magnetic separation, which places an upper

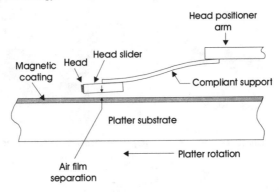

Figure 4.5 *A "flying" head.*

limit on the minimum recording wavelength of the head (see Section 2.2.3.1). Over the years, disk heads have been developed to fly at smaller and smaller separations, driven by the need for ever higher recording density. Today, separations as small as 0.05 to 0.1 μm are common. Because of this flying, disk heads do not face the mechanical stresses of the in-contact heads of helical recorders.

4.1.5.1 Flying Heads

Flying is not a property of the head itself, but that of a carrier, called a *slider*, that supports the head. This is shown in Figure 4.5. As the disk rotates, an air film (rigid disk assemblies are usually sealed in a container with a controlled atmosphere, so the film may not actually be "air") is formed at the interface between the disk surface and the atmosphere. This moving film lifts the head slider slightly off the disk surface to provide the controlled separation. The slider is mounted on a very light compliant arm so it can move up and down a little to follow any vertical runout of the disk surface. With careful design, the result is a very stable flying height.

The head is mounted at one end of the slider where the most stable separation occurs. Because of the dynamics of flying and the compliant support, the head must be extremely small and low in mass. It is an ideal application for a thin-film head (see Section 4.1.5.2).

The flying separation is maintained by the disk rotation. When rotation stops, the head will settle onto the disk surface, and when the disk is restarted, it will continue to contact the surface until there is enough disk speed to reestablish the air film. To control this situation, disk drives are designed so that the head will be moved to a position near the center of the disk whenever the disk stops. A special surface is placed there to support the head while starting or stopping the disk (see Section 3.3.2).

4.1.5.2 Thin-Film Heads

The thin-film fabrication techniques of etching, evaporation, sputtering, and so on, which have been developed so highly in the integrated circuit business, are also adaptable to

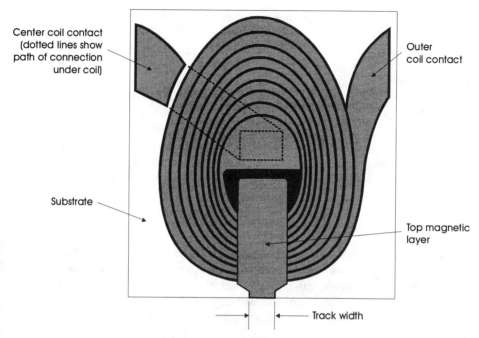

Center coil contact
(dotted lines show
path of connection
under coil)

Outer
coil contact

Substrate

Top magnetic
layer

Track width

Figure 4.6 *Construction of a thin-film head for rigid disk applications.*

head fabrication. These techniques are capable of making circuits or devices that are far smaller than the typical assembled magnetic head. Thin-film techniques are not suitable for building the larger heads, but rigid-disk heads, which now use track widths of 25 μm or less, are well within the range of thin-film methods.

Figure 4.6 shows a typical thin-film head (TFH) construction. A substrate material, usually a ceramic, is the starting point. When the head is completed, the substrate typically becomes the slider of the flying head assembly.

Fabrication begins with an adhesion layer being applied to the substrate to support the first layer of magnetic material, which is applied next. The material is typically a NiFe alloy, deposited about 1 μm thick. This is subsequently etched off at the bottom to determine the track width of the leading pole tip. Next, an insulating layer covers the entire magnetic coating to protect the bottom contact for the coil. The insulating layer over the pole tip area may become the gap spacer, although it is probably better to remove that and deposit a more controlled material for the gap spacer. This insulating layer must have holes etched in it at the center of the coil to expose the magnetic material beneath. This area will become the back gap of the magnetic structure.

Copper is deposited to make the bottom contact of the coil; again, an etched hole in the copper allows the next magnetic layer to reach the first magnetic layer at the back gap area. Another insulating layer now covers the copper, and a hole is etched for the coil layer to contact the bottom contact layer. Now, the coil pattern is made of an etched copper layer, followed by another insulation layer to protect the coil against shorting by the final magnetic layer.

The final magnetic layer is deposited over the region from the center hole in the coil down to the pole tip region of the first magnetic layer. This new layer matches the track width of the first layer, so the active pole tips and gap are now located at the edge of the substrate.

This fabrication scenario is actually done on a large wafer of substrate that can hold hundreds or even thousands of heads. That layer is sliced into bars along the line of pole tips and the pole tips are then lapped to the correct gap depth while in bar form. Lastly, the bars are sliced into individual heads with the substrate slider attached. The process is almost entirely automated and heads are fabricated in very large batches. This is a method to produce high-performance heads at low cost.

4.1.5.3 Magnetoresistive Heads

Although MR read heads deliver more signal output than inductive read heads and they are simple in construction, they suffer from the disadvantage of not being usable for recording. Thus, a separate inductive record head is always needed. In some applications, the advantage in output signal, which translates to being able to read higher-density recordings, is important enough to justify the extra complexity of separate write and read heads.

As explained in Section 4.1.1.2, MR heads must be biased magnetically to obtain response to both polarities of record magnetization. Biasing is the main mechanical complication to the design of an MR head; there are many approaches. Figure 4.7 shows a design using two MR elements separated by a nonmagnetic spacer. A current i_S delivered to the head is divided equally between the two elements. Through each element, the current generates a field that biases the other element; the result is that the elements are biased in opposite directions. This means that when both elements are reading the same recorded flux, there will be no net change in resistance because the resistance changes cancel out. However, when a flux transition occurs between the two MR elements, their resistances both change in the same direction, resulting in a net change in resistance for the parallel elements. The change is maximum when the flux transition is at the center of the head.

A further complication occurs because the iron-based MR element is subject to corrosion. It can be protected on all surfaces except the surface facing the record surface. This is acceptable in the controlled environment of a rigid disk drive but it is not good enough for applications exposed to the air. In that case, some designs have added magnetic pole pieces or flux guides to direct flux from the records to an MR element that can be completely embedded within the head structure. This, of course, makes an MR head cost nearly as much as an inductive head, but it still retains the performance advantage of MR.

4.1.6 Erase Heads

Longitudinal-tape erase heads are designed to have a large gap or multiple gaps. They are driven by a signal to produce a steady high-frequency magnetic field. A large gap makes

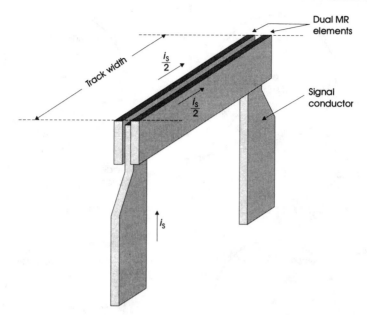

Figure 4.7 *Construction of a magnetoresistive read head.*

it easier to produce the strong field needed for erasing. The erase frequency should be higher than any frequency recorded on the tape. Erasing occurs as the tape passes over the head and then leaves the head. Thus, the erasing field reduces as an area of the tape moves away from the erase head gap, producing the decaying-field effect needed for successful erasing to zero magnetization (see Section 2.2.5). Because of the high-frequency, high-power excitation, heating of erase heads is an important consideration.

Erase heads are used on helical scanners to erase one track or part of a track for editing; these are called *flying* erase heads and they handle erasure differently. Since they erase only one track at a time, and maybe only for a portion of the track, the timing of applying erase current must be carefully controlled relative to the time of applying record current to account for the physical difference between the respective head positions. In helical recorders, this is handled by the edit timing circuits built into the recorder (see Section 8.3.1.5).

The other consideration of flying erase heads is track width. If guard band recording is being used, an erase head would generally be wider than the nominal track width so that a portion of the guard band space is also erased. That assures that tracking error in subsequent replay will not cause the read head to reach into unerased areas. With azimuth recording, the choice of erase head width is even trickier because a too-wide erase head can result in some unrecorded areas being left in the track pattern. Those spaces could cause trouble in the presence of replay tracking errors. With digital recording, it is generally best to make the erase track width the same as the track and expect the digital error protection to take care of any spurious signals from tracking errors.

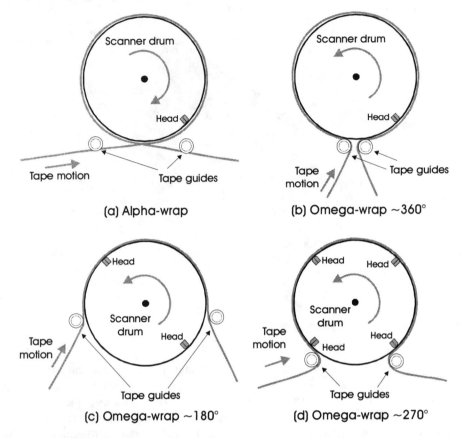

Figure 4.8 *Helical scanning approaches.*

4.2 HELICAL SCANNERS

Some of the considerations of helical tracks were covered in Section 3.2.4. This section discusses the mechanisms for creating helical tracks. All helical scan mechanisms involve wrapping tape around a rotating scanner device. There are many variations of that basic idea.

4.2.1 Basic Formats

Figure 4.8 shows four basic helical scanning approaches. The *alpha-wrap*, Figure 4.8(a), wraps the tape all the way around the scanner so that edges of the outgoing and ingoing tapes pass next to each other. This allows video to be recorded with a single head, although there is a slight gap as the head passes from one edge of the tape to the other. By synchronizing the drum rotation with the video's vertical scanning, this gap can be made to occur within the VBI.

Because essentially all of today's video recorders use cassette tape packaging, alpha-wrap has the serious disadvantage that the tape must be made into a complete loop around the scanner, which does not lend itself to the automatic threading required with cassettes. Thus, it is no longer used.

A second approach for making a one-head video recorder is the full *omega-wrap*, shown in Figure 4.8(b). In this case, the tape wraps only about 350° around the scanner so that a complete loop is not required. This configuration is used in the SMPTE Type C analog video recorders for broadcasting (see Section 10.1.1), which is a reel-to-reel system, but it has not been used in any cassette-based recording systems.

The preferred configuration for nearly all cassette video recorders is the partial omega wrap that uses two heads per channel on the scanner (Figure 4.8(c)). In this case, only a little more than 180° of tape wrap is required and automatic threading is easier to implement.

The two head-per-channel approach requires a drum diameter twice that of the one-head-per-channel approach, which is a disadvantage for size and cost. An alternative technique developed for camcorder applications uses four heads per channel with slightly more than 270° tape wrap on the drum, as shown in Figure 4.8(d). The drum diameter thus can be made nearly as small as the single-head approach.

4.2.2 Scanning Drums

The helical scanning station can be built in several ways, depending on what part of it rotates. Figure 4.9(a) shows a slot configuration, where a thin platter rotates in a slot in the main drum surface. The video heads are mounted on the periphery of the platter. The tape slides across the surfaces of the stationary upper and lower drums. Most early helical scanners were built this way, but it has important disadvantages for volume manufacturing. The split drum is difficult to make because the upper drum must be mounted to a shaft going through the rotating platter. Also, the tape slides across most of the scanner surface, and the motion of the head plate contributes little to an anti-friction air film under the tape.

Modern scanners are all designed with the entire upper drum rotating and carrying the heads on its lower periphery as shown in Figure 4.9(b). This method has fewer precision parts and is simpler to manufacture. The rotation of the upper drum goes a long way to reducing friction of the tape over the scanner, although the transport design must take care not to create any situation where the tape might catch or stick to the rotating drum, either on start-up or while running.

4.2.3 Tape Guiding on Scanners

Tape guidance into and out of a helical scanner must be precisely done to assure that the tape flows smoothly onto the scanner drum and similarly leaves it at the other end. Since the scanner is at an angle to the basic tape path because of the helix angle, the tape must

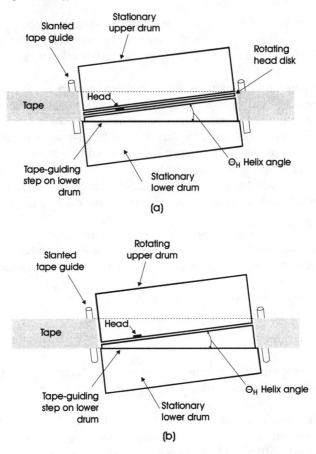

Figure 4.9 *Types of helical scanning drums: (a) head drum rotates in a slot and (b) upper drum rotates.*

twist when coming from the scanner to fit the rest of the tape path. This is usually done with tape guides precisely set at the proper angles (see Section 5.4).

For proper positioning of the helical tracks across the width of the tape, it is usually necessary to locate a tape edge guide going all the way around the scanner, or at least at points around it. This edge guide is machined into the lower drum surface, eliminating adjustments; with modern computer-controlled machining, this is not expensive.

4.2.4 Rotary Transformers

It is necessary to get signals to and from the heads on the rotating member of a helical scanner; this is done with a *rotary transformer*, which has one channel for each head on the drum. This transformer must handle the recording power level and also operate noise-free at the low levels of the playback signal. It poses many design challenges.

4.2.5 Head-to-Tape Contact on Helical Scanners

The rotating drum carries an air film on its surface that could interfere with achieving good head-to-tape contact (small separation). This is overcome by having the head pole tips penetrate above the surface of the drum, causing the tape to be locally deflected by the head and increasing the pressure between tape and head. At the same time, there must be air relief space around the head so the air film over the head can escape. The design of the scanner must also assure that the proper contact conditions are obtained throughout the track length. This is especially critical at the start and end of the track, where the dynamics of the head entering and leaving the tape can cause local disturbances of contact.

4.3 OPTICAL HEADS

The basic concept of an optical read head was shown in Figure 2.9. This section covers the concept in more detail.

4.3.1 CD and CD-ROM

The design of read heads for both CD and CD-ROM uses the concept described above but must be packaged in a form that is suitable for low-cost manufacture and a compact drive package. A typical design is shown in Figure 4.10. The electromechanical actuator for moving the objective lens to focus and track the recording is designed with two sets of magnets and coils to obtain the two-axis movement required. This is mounted to an optical substrate that supports the other components behind the actuator, as shown. To bring the optical path onto the substrate, a right-angle prism bends the optical path between the objective and the quarter-wave plate. Of course, this entire assembly rides on an arm or carrier that delivers the radial motion needed for track selection anywhere on the disc. The major components of a CD read head are as follows:

- *Laser*—This is a GaAlAs (gallium-aluminum arsenide) infrared semiconductor diode laser with a wavelength from 0.78 to 0.83 µm. It has a power output of about 3 mW.

- *Objective lens*—This is a microscope objective type of lens whose performance can be described by its *numerical aperture* (NA), which is the sine of the half-angle for the cone of light passing through the lens. The approximate diameter d_S of the spot of light produced by a lens is equal to:

$$d_S = 0.82 \, \lambda \, / \, NA \qquad (4.2)$$

CD read heads use a lens of NA = 0.45, which gives a spot size of about 1.5 µm (60 µin) at the record surface within the disc. Because the 1.2-mm thickness of the transparent disc substrate is between the lens and the record surface, the light spot is grossly out of focus at the disc surface, preventing all but the most major surface contamination or defects from affecting the readout. The lens NA of 0.45 is a compromise between the requirement for a small spot and the optical depth of focus and alignment tolerances.

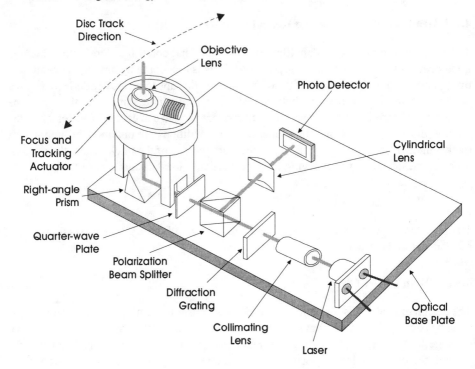

Figure 4.10 *Construction of a CD-ROM read head.*

- *Photodetector*—This is a semiconductor diode of the *pin* (positive-intrinsic-negative) type. As was shown in Figure 2.9, it is divided into segments to support focus and tracking detection as well as signal detection.
- *Collimating lens*—This assures that all the light output of the laser is directed into the optical path.
- *Diffraction grating*—This splits the laser beam into a main spot and the two side spots needed for tracking servo control.
- *Quarter-wave plate*—This provides the optical phase shift needed to direct the return beam reflected from the disc off to the photodetector.
- *Polarizing beam splitter*—This passes the laser light directly through to the objective lens but it directs the return beam off to the photodetector because of the phase shift caused by the quarter-wave plate.
- *Cylindrical lens*—This spreads the three spots apart on the surface of the photodetector.

Servo electronics for CD readers are described in Section 6.2, and signal electronics are described in Section 11.1.

4.3.2 CD-R and CD-RW

The basic optical path of a CD-R or RW drive is the same as for CD-ROM except that focus and tracking work differently during recording. Before recording, there are no pits on the disc; thus, to provide a means for focus and tracking, grooves are pressed into the surface of the blank disc, as was shown in Figure 3.7. Recording takes place at the bottom of the groove. During recording, a three-spot servo senses the edges of the groove and controls tracking to keep the main recording spot centered at the bottom of the groove. Focusing is controlled by observing the asymmetry of the main spot, the same as in a CD reader.

Recording occurs by deformation of the dye layer when the recording laser is switched to high power. On replay, the deformation affects the reflectance seen by a reader and may be read with the same head used for reading conventional pressed CD-ROM discs. Between the recorded areas, which correspond to the pits of a normal CD, the laser cannot be switched completely off because that would interfere with the tracking operation. However, it can be set at a power level suitable for tracking but not high enough to perform recording.

4.3.3 DVD Heads

As already explained in Section 3.4.2.2, DVD provides a four times improvement in area density compared to CD by the use of shorter wavelength light (0.65 µm) and closer track spacing. That requires a smaller spot diameter, which is obtained by using an objective with a higher NA (0.8) along with the shorter wavelength light. From (4.2) the spot diameter is thus about 0.7 µm.

Most DVD-ROM readers are designed to also read standard CD-ROM discs and sometimes CD-R discs. This requires focusing to different distances (CD focus distance is about 1.15 mm and DVD focus distance is 0.55 mm). Various approaches have been used for this, including special holographic lenses that can focus at two distances, or twin lenses, and even twin lasers.

In a DVD system with two data layers per side, the layers are selected by changing the focus distance—the distance between layers is about 0.05 µm. When a disc is first read, the layer closest to the read head is acquired. The second data layer is acquired by shifting the focus deeper into the disc and enabling the focus servo. The data format itself tells the reader which layer is being read.

4.4 CONCLUSION

Heads and scanners do the work of a video recorder. The rest of the system is concerned with presenting the medium to the heads and scanner, and processing signals for them.

REFERENCE

[1] Jorgensen, F., *The Complete Handbook of Magnetic Recording*, 4th Ed., McGraw-Hill, New York, 1995.

5

Tape Decks

The main mechanical assembly of a tape recorder is a *tape deck*. Its function is to hold the tape supply, usually a cassette, and manipulate the tape to present it to the heads and scanner. Additionally, it must be able to search forward in the tape or rewind it back to the starting position when recording or playing is finished.

Essentially all videotape recorders are of the helical-scan type. The only longitudinal tape recorders used for video are digital decks that are principally used to archive digital video (and any other type of data) on computers, but may not play video in real time. Thus, this chapter covers only helical-scan tape decks.

Helical scan tape decks come in all sizes and shapes, from rack-mounted and desktop to the decks built into camcorders. Early video recorders had open reels and the tape had to be placed in the tape path (*threaded*) by hand. Today all use tape in cassettes and threading is automatic. The focus here is on that type of system.

5.1 THE TYPICAL TAPE DECK

All tape decks have a set of components to accomplish the various tasks of the deck. Figure 5.1 shows a typical tape deck and the types of components that are involved. These are summarized here.

- *Baseplate*—This supports all the parts of a tape deck, and is usually a metal casting, although fabricated baseplates are sometimes used in the small decks for camcorders.

- *Tape cassette*—All modern videotape recorders use a cassette to hold the tape when it is outside the tape deck. The most common format has two reels side-by-side that engage with drive mechanisms when the cassette is inserted into the deck. (See Section 5.6.)

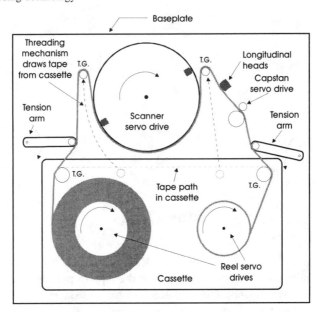

T.G. = Tape guide

Figure 5.1 *A typical video cassette tape deck.*

- *Tape threading mechanism*—Except for audio tape decks that have no helical scanner, all cassette decks have a threading mechanism to draw the tape from the cassette and place it in the deck's tape path. (See Section 5.7.)
- *Reel drive servos*—Most tape decks have servomechanisms for the reel drives to control tape tension under all conditions of operation. This is necessary for proper control of the tape in the deck and also to standardize the tape tension at the helical scanner. (See Section 5.5.1.)
- *Tension arms*—These are spring-loaded rotating arms with a post or roller running against the tape. They have two purposes: (1) to provide a degree of compliance in the tape path other than the compliance of the tape itself, which assures that the tape is not stretched excessively under dynamic conditions of changing tape speed modes; and (2) to provide means for sensing the tape tension for servo control purposes. (See Section 5.3.4.)
- *Tape guides*—These are necessary to change the direction or angle of the tape path, and to control the height of the tape path above the tape deck baseplate (*edge guiding*). (See Section 5.3.2.)
- *Helical scanner*—The helical scanner accommodates the helix angle into the tape path. (See Section 5.4.1.)
- *Helical drum servo*—Since tape record specifications require that helical tracks be synchronized with video field or frame scanning, a servo drive is required to control the scanner drum speed and phase. (See Section 5.5.3.)

- *Longitudinal track heads*—These heads are mounted separately at a specified point in the tape path to meet the record pattern standards.

- *Capstan drive*—Tape speed is controlled by a rotating capstan; the tape is generally held in contact with the capstan by a compliant pinch roller. (See Section 5.3.5.)

- *Capstan Servo*—For precise control of tape speed during recording and for control of tracking during replay, a servo system is required on the capstan. (See Section 5.5.2.)

Different tape decks combine these components in different ways; certain decks even eliminate some of them. The components and their design considerations are discussed in the rest of this chapter. It is important, however, first to discuss the tape itself and how it affects tape deck design.

5.2 PROPERTIES OF TAPE

The mechanical properties of tape are well known, but their implications to tape deck design are reviewed here. Tape is flexible and elastic; it must be kept under tension at all times to control its position. However, this stretches the tape, changing its dimensions; too much tension can cause tape damage or even breakage.

Because tape is so thin, very little force can be applied to its edges; too much force and it curls over, which results in loss of control and possible edge damage. The recording coating of tape must by necessity be very thin because of magnetic considerations. At the same time, it often has to run over tape guiding components; these components must be carefully designed and manufactured to assure that they do not damage the tape surface.

Depending on the tape surface properties, the deck components, and the environment (especially humidity), tape may sometimes stick in the tape path and not move when required. This problem is minimized by selection of materials, control of tape tension to prevent it from becoming too high, and by simply avoiding extreme humidity conditions. Some tape decks contain a moisture sensor that will issue a warning or prevent operation entirely when the humidity becomes too high.

Finally, tape is rolled on reels. This process must be carefully controlled to prevent reels or their flanges from causing the potential damaging effects mentioned above. The remainder of this section discusses some of the design considerations for tape decks that arise because of tape properties.

5.2.1 Tension

Tape tension must be present under all conditions of tape deck operation; if it is lost, tape becomes uncontrollable. Furthermore, tension must be held in a certain range to control tape stretching, tape motion, and to avoid tape damage. As tape moves through a deck for recording or replay, tension is achieved by holding back on the supply reel with a brake or motor. Additional tension is produced by friction drag caused by the tape moving over the other elements in the tape path. Thus, in forward motion of the tape, tension is least at the supply reel and increases at each friction point downstream in the tape path. This accumulation stops at the capstan, which clamps the tape and exerts whatever force necessary to move the tape at the correct speed.

Downstream from the capstan, tension is controlled by the takeup drive. The actual load on the capstan motor depends on the difference between the tension on the supply side of the capstan and the tension on the takeup side of the capstan. The capstan loading can be either positive or negative; the clamping of the tape to the capstan by the pinch roller must be good enough to hold the tape against the capstan under all these conditions.

When tape is being rewound or fast forwarded, most transports disengage the capstan pinch roller and tape speed is controlled entirely by the reel drives. These are controlled to deliver the proper forces needed to move tape at high speed through that particular deck. However, the speed is quite variable when operating this way.

Some decks can also operate at controlled slow or fast tape speeds for slow- or fast-motion replay. The capstan is engaged for these modes to achieve precise tape speeds.

In recording or replay modes, both reel drives must deliver a specific tension value, independent of the amount of tape that may be on either reel. Although early longitudinal audio tape decks often achieved tension control with mechanical brakes for holding back, and a single drive motor with clutches to control tape direction, these methods may not be precise enough for a helical-scan video deck. Professional helical decks have a motor for each reel. Each motor is controlled by a servomechanism that responds to tension sensors (usually tension arms, see Section 5.3.4) in the tape path. Home helical decks use a single motor at the takeup with a friction holdback controlled by a tension arm in a mechanical servo arrangement (see Sections 5.3.6 and 5.5.1), or they may take power from the capstan motor. On rewind in home systems, a clutch transfers the takeup motor drive to the supply reel shaft.

5.2.2 Tape Edge Guiding

Control of tape positioning in the direction across the width of the tape is achieved by various edge guiding means. As already mentioned, the amount of force that can be exerted on the edges of tape is very limited, which means that the transport design must achieve its tape path almost without any edge guiding. This requires precise alignment of all tape path components. However, a slight amount of edge guiding is still necessary to take up small variations in alignment or tape elastic properties. Edge guiding posts or rollers are used for this purpose; the tape is always wrapped around the guide through a small angle to make a cylindrical segment that will be stiffer at the edges than a flat tape segment would be (see Section 5.3.2).

5.2.3 Tape Reels

Tape reels that are used by themselves (without a cassette) must have flanges to protect the tape pack from disturbance by manual handling. Flanges also play a part in the dynamics of fast loading the tape onto a reel because the air escaping from between the layers of tape provides a guiding effect. However, the tape guidance into the reel must still be accurate enough that the edges of the tape do not touch the reel surface when entering the reel. A properly operating reel system will load tape onto the reel without the edges of the tape pack touching the flanges at all.

In these days of tape cassettes, flanged reels are generally used inside the cassette, although a few cassette designs have only hubs—the cassette body provides the flange function. However, the same principles of guidance apply: the tape should be guided into the cassette without its edges touching any part of the cassette or its reel flanges. Any scraping of the tape edges across the cassette surface will cause disturbance to the tension control servos and can result in unsatisfactory performance and possible tape damage.

5.3 TAPE DECK COMPONENTS

The previous sections discussed the purpose for each tape deck component. This section discusses design considerations for each component.

5.3.1 The Baseplate

This is the support for all the tape deck components. It provides long term stability of position and alignment for all the critical components. Many decks use a die-cast metal baseplate that is designed to provide rigidity and stability at all necessary mounting points. Castings are heat-treated to relieve any internal stresses, but the resulting base plate usually will not have the necessary accuracy at critical mounting points. Thus, these mountings are machined after casting to achieve the necessary accuracy because that is cheaper and more reliable than making the mountings adjustable during assembly.

Light-weight decks for camcorder applications often have a fabricated baseplate rather than a casting. Because of the small overall dimensions of such decks, acceptable rigidity can be obtained with surprisingly thin metal parts.

5.3.2 Tape Guides

Any component of a deck that the tape wraps around, even slightly, is a *tape guide*. It is capable of changing the tape path direction, height above the baseplate, or angle with respect to the baseplate. Figure 5.2 shows some examples of tape guides.

For tape guiding to be effective, the tape must wrap around the guide, as shown in Figure 5.2(a) for a perpendicular guide. The wrap angle θ_W is the angle on the post between the tangent points of the tape entering and leaving the guide, which is also the angle by which the direction of the tape path is changed. Of course, larger wrap angles cause more friction drag as the tape goes around the guide post. Some decks use rotating tape guides (rollers) where the post surface rotates with tape movement. This reduces friction drag but there is a trade-off because a rotating guide is more expensive and may cause slippage during rapid tape speed changes, which could cause tape damage.

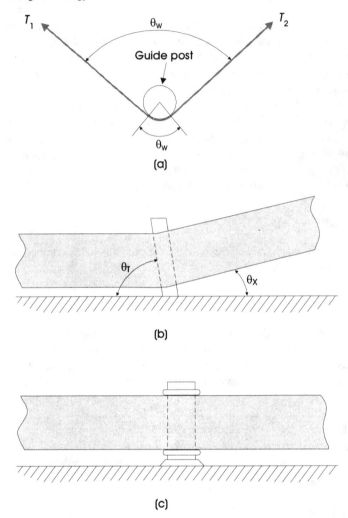

Figure 5.2 *Some examples of tape guides: (a) perpendicular guide (top view), (b) tilted guide (side view), and (c) edge guide (side view).*

The tape path height above the baseplate is changed by using a tilted tape guide as shown in Figure 5.2(b). Generally, two tilted guides are used together to raise or lower the tape path and bring it back to a level condition. The angle θ_X that the tape path makes with the baseplate from a guide tilted at an angle θ_T in the direction of tape motion depends on both that angle and the wrap angle θ_W. A tilted tape guide may generate a lateral force on the tape, which tends to move it in the direction of the tilt. This must be counteracted with proper edge guiding (see Section 5.3.2.1).

5.3.2.1 Tape Edge Guiding

Edge guiding is achieved by adding an extension on a guide post at one or both edges of the tape, as shown as in Figure 5.2(c). As mentioned before, the tape should have some wrap on the post to stiffen it to prevent edge curl at the guide. When both edges are guided at the same post, a clearance must be allowed to accommodate the widest possible tape. This is less critical when the two tape edges are guided at different posts because the compliance of the tape web between the posts will take up the tape width tolerances. Guide wear is another consideration because of the narrow tape edge riding continuously at the same point on the edge guide; materials must be carefully chosen for this.

5.3.3 Mounting of Longitudinal Heads

All items on the tape path must be carefully aligned to achieve proper running of the tape under every condition of operation. However, longitudinal heads have the additional requirement of being positioned for recording or playing the proper track patterns. This requires setting the height of the heads to control the position of the track from the edge of the tape, as well as angular positioning to achieve the correct head azimuth. Considering that factors such as tape edge guiding can affect the height of the tape itself in the path, longitudinal head mountings often require mechanical adjustment to take up accumulated tolerances. Adjustments are normally a part of the manufacturing process and are done either with a series of screws or, in low cost products, by bending the mounting at points designed for that purpose.

Adjustment of track height and azimuth for longitudinal heads can be done by playing a test tape recording known to have standard tracks and adjusting both head height and azimuth for maximum output. Even finer adjustment of height is done by making a new recording with the head under test and developing the magnetic track for measurement of position using a microscope (see Section 3.2.5).

It should be pointed out that proper head-to-tape contact on longitudinal heads generally requires a specified tape wrap angle on these heads. This is provided for in the tape path layout by suitable positioning of guides.

5.3.4 Tension Arms

Tension arms provide the two functions of sensing tape tension and adding compliance to the tape path beyond that inherent in the tape's elasticity. They usually consist of a post mounted on a swinging arm having a spring that presses the post against the tape, as shown in Figure 5.3(a). A position sensor is coupled to the arm to provide an electrical signal for servo control of tension. Many variations of this basic approach are used on different tape decks, with different arm configurations, different springs, or different tape path layouts.

The tape tension is converted to a force against the arm post as shown in Figure 5.3(b). If one assumes the tape path is symmetrical about the tension arm post and the friction drag on the tape at the tension arm post is negligible (which is a fair approximation for

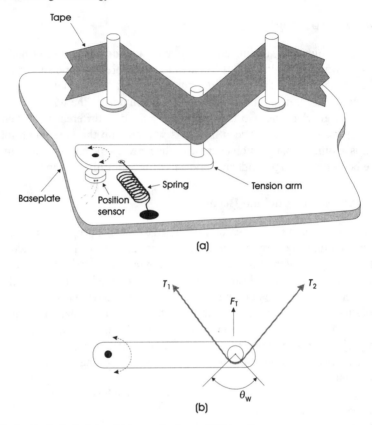

Figure 5.3 *A typical tension arm: (a) isometric view and (b) top view.*

most tape decks), the force F_T that the tape exerts on the tension arm is a function of the wrap angle θ_W:

$$F_T = 2T \sin(\theta_W/2) \tag{5.1}$$

where T is the tape tension force.

Thus, the greater the wrap angle on the tension post, the greater tension force is transmitted to the post. Large wrap angles are difficult to achieve and they tend to increase the friction drag anyway, so tension arm layouts usually have wraps less than 90°. This is not a problem in tension servo design (see Section 5.5.1) because it can be made up with electronic gain, but small wrap has little tape path compliance.

The tension arm swings as tape tension changes because of the action of the force against the arm spring. That causes the wrap angle to reduce with increased tension, creating a nonlinearity. This also is generally not a problem to tension servos because they are designed to operate at a fixed tension value anyway or, at least, in a narrow range of tensions.

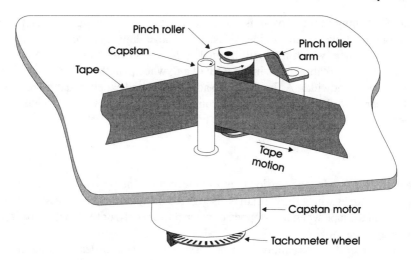

Figure 5.4 A typical capstan drive.

5.3.5 Capstan Drives

The capstan drive, in combination with its servo electronics, provides the precise tape speed control needed in all videotape recording systems. A typical drive is shown in Figure 5.4. The capstan is a precision-ground shaft with a carefully roughened surface that contacts the tape, usually on the magnetic coating side, when the pinch roller is activated to press the tape against the capstan.

The pinch roller is made from a compliant material, such as neoprene, and is wider than the tape width; it generally runs on the back side of the tape. The pinch roller has an activating mechanism that allows it to be released from the capstan when the tape is stopped or is not under speed servo control, as in rewind or fast forward modes. The pinch roller mounting is usually flexible so as to align itself to the capstan shaft and thus will not be an additional critical item of tape path alignment.

A few recorders use a capstan without a pinch roller. In this case, the capstan drives the tape at all times. This requires a larger capstan diameter to allow for the high tape speeds in shuttle mode without the capstan rotational speed becoming excessive. A larger wrap on the capstan is also necessary; 180° or more is common. In earlier days, vacuum was applied inside the capstan to keep the tape in contact; this is extremely expensive and is no longer used for video decks.

The capstan is direct-driven from a direct-current (DC) motor in most video recorders. That avoids any other mechanical drive components and gives the best servo performance. A tachometer wheel is usually provided on the capstan motor to support velocity feedback in the capstan servo (see Section 5.5.2). The capstan motor must run smoothly with minimal cogging or bearing noise, because that would be imparted directly to the tape and could have frequency components outside the bandwidth of the servo loop. Because of this, many capstan motors, even in high-priced systems, do not use ball bearings.

Smooth running of the capstan is particularly important if analog longitudinal audio tracks are used. However, most video decks today have digital audio, usually combined with the video track, which may still be analog. Smoother running of a capstan can be obtained by adding rotating mass in the form of a flywheel, but this is a trade-off because it reduces servo bandwidth. A flywheel also adds weight and expense to the deck because it must be carefully balanced to avoid becoming a source of instability. Thus, capstan flywheels are seldom seen.

5.3.6 Reel Drives

The reel drives provide tape tension control in recording and playing modes and control tape speed in modes where the capstan is disengaged. In addition, reel drives must include means to hold the tape when it is stopped or when power is off, or to stop the tape if power fails with the tape moving.

A typical reel drive assembly for a professional video recorder is shown in Figure 5.5(a). The reel platter is directly driven by a motor that is controlled by a servo amplifier. The reel platter is engaged with the reel of a cassette by the threading sequence of the tape deck (see Section 5.7). A mechanical brake at the back of the motor allows for stopping or holding the tape when motor power is removed. A *tachometer wheel* is at the back of the motor. This is a transparent wheel coupled to the reel shaft and has an integral number of opaque segments around its periphery. An optical sensor with a light-emitting diode (LED) and a photodiode views the wheel and generates a signal whose frequency equals the reel shaft speed multiplied by the number of opaque segments on the wheel.

Home video recorders generally do not have a motor for each reel. Some home decks use the capstan motor as the power source for all reel operations. One such system is shown in Figure 5.5(b). A belt from the capstan motor drives a clutch wheel that can be coupled to either the takeup reel shaft for play or fast forward, or to the supply reel shaft for rewind. Friction drag assemblies on both reel shafts are engaged as required to provide holdback tension in either tape direction. In the no-power condition, both drag assemblies are engaged. The operation of this system requires proper cooordination of control for the capstan motor, pinch roller, clutch wheel, and reel drag brakes. Table 5.1 shows how this is done for the five basic tape deck modes. Engagement of the pinch roller puts the tape speed under control of the capstan servo, and the reel drive provides hold-back and

Table 5.1
Reel Drive Modes for a Camcorder

Mode	Capstan pinch roller	Drag brakes S	T	Clutch wheel
Record	Engaged	On	Off	Coupled to T
Replay	Engaged	On	Off	Coupled to T
Fast forward	Disengaged	On	Off	Coupled to T
Rewind	Disengaged	Off	On	Coupled to S
Stop	Disengaged	On	On	Off

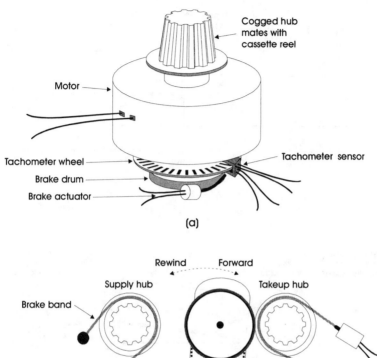

Cogged hub
mates with
cassette reel

Motor

Tachometer sensor

Tachometer wheel

Brake drum

Brake actuator

(a)

Rewind   Forward

Supply hub   Takeup hub

Brake band

Brake
actuator

Clutch wheel

Clutch wheel
arm

Drive belt
from capstan
motor

(b)

Figure 5.5 *Reel drives: (a) professional and (b) home.*

take-up tension.When the pinch roller is disengaged, tape direction is controlled through the position of the clutch wheel and tape speed is controlled by the capstan motor.

5.4 INTERFACING TO A SCANNER

Because of the helix angle, the tape climbs up (or down) the scanner as it goes around the drum. At the cassette, both reels of tape are in the same plane, so the change of height

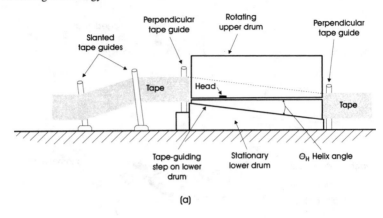

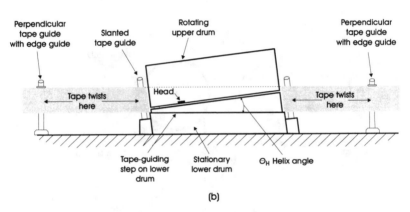

Figure 5.6 *Interfacing the scanner to the tape path: (a) mounted perpendicular and (b) mounted at an angle to the baseplate.*

when crossing the scanner must be accommodated in the tape path between cassette and scanner. There are two basic ways to do this, depending on whether the scanner is mounted perpendicular to the baseplate or mounted at an angle to the baseplate.

5.4.1 Perpendicular Mounting of the Scanner

When the scanner is mounted perpendicularly, the tape surface entering and leaving the scanner can remain perpendicular to the base plate but it changes height above the baseplate when passing the scanner. This can be accommodated by using two tilted tape guides to change the height of the tape path as shown in Figure 5.6(a). The shifting of tape level may be done on one side of the scanner as shown in the figure, or it may be done on both sides.

5.4.2 Tilting the Scanner

Mounting the scanner at an angle to the baseplate, as shown in Figure 5.6(b), requires no height change in the tape path, but it still needs tilted guides at the scanner to twist the tape to the proper angle to meet and leave the scanner. This is the preferred approach in the smaller-format tape decks because it can be accomplished with a shorter total tape path length.

5.5 SERVOMECHANISMS

Most video recorders have servo control for tape speed, scanner speed, tape tension, and tracking. A servomechanism is simply a feedback loop that senses the parameter under control, compares it to the desired value, and amplifies the difference signal to provide control to the actuator or motor in a direction that will reduce the difference signal (negative feedback). The mechanical components involved have already been discussed. Each of the video recorder servo systems is described in this section.

5.5.1 Digital Servo Techniques

In this day of digital integrated circuits (ICs), the preferred approach to servomechanisms is digital, used in nearly all video recorders. In most cases, a special purpose microcomputer is included in the recorder (called an *embedded* microcomputer) and performs the logical processes for system control and all servos. That is easily accomplished because the microcomputer can cycle through each servo system's calculations hundreds or even thousands of times per second. Of course, the microcomputer requires power amplifiers (PAs) to drive motors, and switch drivers to drive brake solenoids and other actuators.

Thus, all servos are controlled by software running on the system's microcomputer, and the hardware appears deceptively simple. The descriptions below do not attempt to detail the software; they describe the servo actions only in functional terms.

5.5.2 Tape Tension Servos

Sensing for a tension servo is a tension arm riding on the tape. The signal from the tension arm position sensor is the input to the servo. The actuator may be either a motor or a friction drag device. Figure 5.7 is a block diagram of a typical supply and takeup reel tension servo system for a professional video recorder. The basic servo operation attempts to maintain a constant output of the tension arm position sensor, which may be a potentiometer or an optical device. In either case, the tension arm signal is digitized and fed to the microcomputer, where it is compared to a reference value and the difference is amplified to drive the associated reel motor.

Most of the complexity of the tension servo is caused by the different modes of operation, listed in the sections below.

5.5.2.1 Tension in Record or Play Modes

In these modes, the supply tension servo provides a constant hold-back tension, while the capstan drives the tape at constant speed. The takeup reel servo is operating similarly, holding the tape tension constant while taking it up on the reel. Both servos thus have a constant reference value.

5.5.2.2 Tension in Fast Forward Mode

In this mode, the supply tension servo continues to provide a constant hold-back tension, but the capstan is now disengaged and the takeup tension servo controls the tape speed. This method of operation allows higher tape speeds than the capstan could provide.

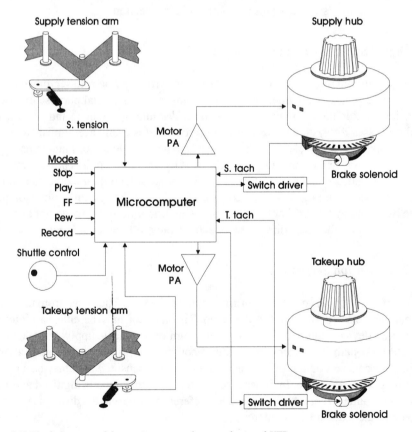

Figure 5.7 *Block diagram of the tension servos for a professional VTR.*

5.5.2.3 Tension in Rewind Mode

In this mode, the supply tension servo controls the tape speed, while the takeup servo provides hold-back tension; the capstan is disengaged. Setting a higher reference voltage on the servo sets the supply tension higher than the takeup tension, causing the tape to rewind. However, this simple system does not control tape speed very well, especially as the tape pack moves from one reel to the other.

Better speed control is obtained by using the tachometer wheels on both reels to sense actual tape speed. A speed profile can be developed by comparing the frequency signals from both reel tachometers to calculate how much tape is on each reel and setting the speed reference voltage accordingly. The microcomputer easily handles this.

Some tape decks have a variable-speed *shuttle* mode for tape winding, where a single knob controls tape speed either forward or backward. This is very useful when searching a tape for a specific location, especially when the video system allows viewing of the picture during this operation (*picture-in-shuttle*—see Section 9.7.1). Shuttle requires control of the tape winding mode of both servos from the single control, which can be accomplished digitally by dynamically computing speed and tension profiles for both reel servos.

5.5.2.4 Tension in Stop Mode

There are a number of situations when stopping the tape is necessary.

- When the tape is stopped from record or play modes, the capstan controls the stopping and the reel servos simply continue in their constant-tension mode. This maintains tape tension around the helical scanner, so a still-frame picture may be displayed in this condition if the tape format and signal electronics allow it.

  The scanner, however, is continuously scanning the same section of tape in a still-picture mode, which can cause tape damage if it lasts too long. Because of this, most decks have a timer that will relax the tape tension on the scanner after the tape remains stopped for a specified amount of time. If the tape is held by the capstan, this can be done by reducing the supply tension to a low value. (It cannot be reduced to zero because the action of the scanner might begin moving the tape.) Another approach to prevent tape wear in still-picture mode is to periodically move the tape to a new position, so the still picture changes to an adjacent frame every few seconds whether or not the operator wants that.

- Stopping from fast forward or rewind modes requires programming the tape speed to zero with the reel servos and then either applying the mechanical brakes or engaging the capstan to hold the tape. Either of these actions must be carefully timed to prevent tension surges.

- The tape must be stopped when approaching the end (or start, in rewind) of the cassette. The tape ends are tightly fixed to the reels in the cassette, so running into the end of the tape, even at play speed, can cause a tension surge that might produce tape or head damage. That means the system must somehow anticipate the approach of a tape end and suitably slow down before the end is reached. In play mode, this can be done by sensing the supply reel rotation speed using a reel shaft tachometer wheel; since the tape speed is constant, the reel speed increases toward the end because of reducing diameter. However, most cassette designs provide a simpler method: a section of tape at each end of the cassette pack is made transparent and an approaching end can be detected optically to initiate slowdown.

The detection of tape ends is more difficult during shuttle mode, because the exact tape speed is variable and unknown, but the optical sensing method will still work.

- The reel servo system should detect fault conditions in the tape path. For example, if the tape sticks in the path and does not move when it should, the system should shut down and signal a fault to the user. This is another use for the reel tachometers, which will show an indication that the tape is broken or not moving if either reel shaft fails to rotate when it should.

- Finally, the tape must be stopped when power is turned off or lost while tape is running. This is normally a function of the mechanical reel brakes, which are designed so they will engage upon power loss.

5.5.2.5 Winding Tape onto Reels

The process of winding tape onto reels is more complex than it seems, especially when it is done at the high speeds that are required for rewinding or shuttling. Tape moving at high speed carries an air film with it, which is squeezed out as the layers of tape build on a reel; this action depends on both tape tension and speed. There must be sufficient clearance between the tape pack and reel flanges for the air to escape. Careful control is required to maintain good "packing" of tape on the reels. A winding tension surge will cause the reel packing to be disturbed, and a slight mechanical misalignment may cause the tape to momentarily touch the reel flanges during a tension disturbance. These effects cause an uneven pack build-up, which may cause disturbances in the tape deck the next time that tape is used.

5.5.3 Capstan Servos

The purpose of the capstan servo is to control tape speed precisely during recording so that a correct record pattern is created, and to control replay tape speed so that proper tracking of the recorded video tracks is achieved. In some tape decks, the capstan motor also runs the reels, as described in Section 5.3.6.

A typical capstan servo block diagram is shown in Figure 5.8. In record mode, the capstan is run at a precise rotational speed by comparing the frequency generated by the capstan's tachometer wheel to a precise reference frequency. Since most tape record standards require the recorded pattern to be synchronized with some features of the video scanning, such as the vertical blanking interval, the capstan reference frequency is usually derived from the incoming video signal. The exact frequency required depends on the

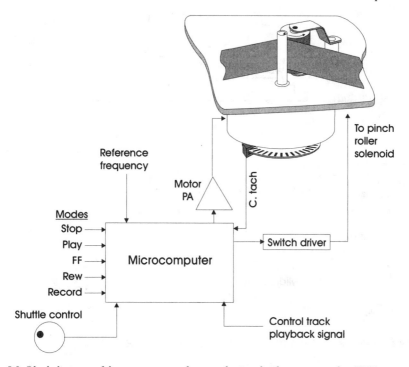

Figure 5.8 *Block diagram of the capstan servo for a professional videotape recorder (VTR).*

tape speed standard, the capstan shaft diameter, and the number of segments on the capstan tachometer wheel. These elements are designed so the required reference frequency bears a convenient integral relationship to the video scanning frequency.

In replay mode, the capstan is controlled by a tracking signal, which is usually received by replay of the control track on the tape. A phase-lock loop (PLL) is made by comparing the control track signal to a reference signal derived from the video system that will receive the output from the recorder. The servo operates to bring the two pulses into coincidence. In a PLL servo, feedback from the capstan tachometer wheel can also be used to help stabilize the capstan operation.

5.5.4 Scanner Drum Servos

The scanner drum also must be phase-locked to the incoming video during recording to assure the correct record pattern. During playback, the same phase-lock loop is used to lock the scanner to the video reference signals. This is shown in Figure 5.9.

For scanners that make one track per video field, a tachometer wheel generates a pulse corresponding to the time of each head beginning a track that is phase-locked to the field-scanning pulse of the video system. For segmented helical video track patterns, the once-per-head pulse from the scanner is phase-locked to a signal generated from the video input at the correct segmentation frequency and phase.

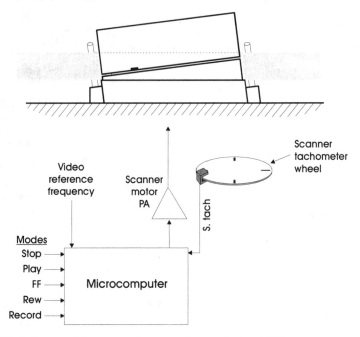

Figure 5.9 *Block diagram of the scanner drum servo for a professional VTR.*

5.5.5 Head Tracking Servos

The basics of head tracking from a control track signal were discussed in Section 5.5.2. However, this type of servo does not actually respond to the video tracks themselves; rather, they set up a fixed relationship to the video tracks. Such servos still require manual adjustment to match the servo relationship to the actual tracks on tape. A different method of control allows optimal tracking to be obtained automatically and even has the possibility of following tracks that are recorded at an incorrect helix angle or that are not straight, or to play back at different tape speeds for slow- or fast-motion effects.

Of course, a multihead approach could be used to track a magnetic recording by having two extra heads to read at both sides of the desired track like optical systems do (see Section 2.3.1.1). That is mechanically complex and expensive, however, and requires more heads on the drum. Most systems use a single replay head that is mounted so it can be moved in a direction perpendicular to the rotation of the scanner drum. The most common approach uses the piezoelectric bimorph mounting, shown in Figure 5.10. Each bimorph arm has two layers—a piezo layer and a metal support. When voltage is applied to the piezo element, it shrinks or expands and causes the bimorph arm to bend. The dual mounting assures that the head penetration will not change as the head is moved up or down. Strain gages on one bimorph arm are used to sense the arm's position for servo feedback.

Control of this tracking actuator is achieved by sensing when the modulated video

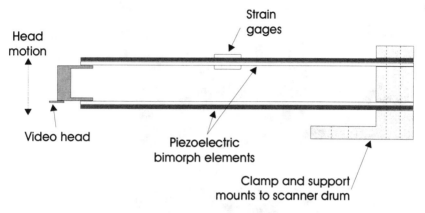

Figure 5.10 *Bimorph head mounting for automatic tracking system.*

signal played from the tape is at maximum signal output, which corresponds to the head exactly following the recorded track. As the head moves off either side of the desired track, output will reduce. This is a *maximizing servo*; it requires the use of *dither* to produce an error signal that will go to zero when the output is maximum. Figure 5.11 shows how this is done.

A small signal of approximately 500 Hz is applied to the bimorph drive to cause it to wander (dither) from side to side as tracks are being read. The modulated carrier output from the head is detected and the phase of the dither modulation on the signal is observed. As the figure shows, the dither waveform is full-wave rectified when on-track; the dither modulation becomes unbalanced when off-track to either side (the phase of the unbalance indicates which side is off). Sampling the carrier signal amplitude at two points one-half cycle apart at the dither frequency and taking the difference produces an error signal. The servo system adjusts the average bimorph position to move the error signal toward zero. Such a servo can have a bandwidth up to 1/2 the dither frequency, which even allows a track that is not straight across the tape to be correctly followed.

5.6 CASSETTES

All modern videotape decks use a cassette to hold the tape for protection during handling and to allow automatic threading. The tape is captive on two reels within the cassette, but, when inserted into a deck, a door opens to allow the tape deck's threading mechanism (see Section 5.7) to draw the tape out into the tape path and wrap it around a scanner for recording or replay. Some typical cassettes are shown in Figure 5.12. The considerations of cassette design are discussed below.

Figure 5.13 shows a disassembled VHS cassette. This cassette includes most of the features of all cassettes. A two-piece plastic body holds all the components. The features include the following:

Signal waveforms modulated by dither

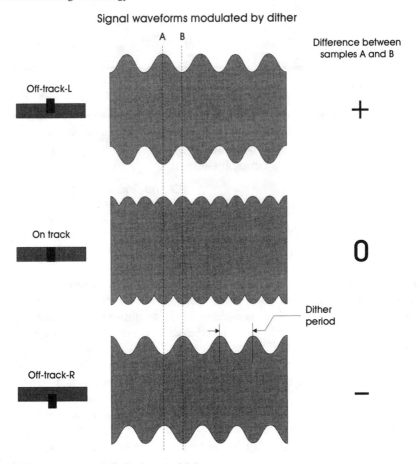

Figure 5.11 *Automatic tracking by the use of dither.*

- *Tape reels*—The tape path in the cassette moves from the supply reel (top in the figure) out along the long edge of the cassette, where it can be accessed by the tape deck threading mechanism. Tape in video cassettes is always B-wrap—with the magnetic coating facing out; this is required by the need to wrap the tape around a helical scanner.
- *Tape door*—This is seen at the right side of the top cover in the figure. It covers the tape going along the edge of the cassette when the cassette is outside of the tape deck. The door is automatically opened when the cassette is inserted into the deck.
- *Tape threading access*—The areas behind the tape that can be seen in the bottom cover provide access for the threading pins and guides to enter the cassette as it is loaded.
- *Reel locking*—Springs in the top cover press the reels down against the bottom

Figure 5.12 Typical tape cassettes.

cover of the cassette when it is not in the deck. This serves the dual purpose of sealing the opening against dust and locking the reels against inadvertent turning, which would loosen the tape in the cassette.

- *Reel drive access*—Holes in the bottom cover allow the deck's reel drives to couple to the cassette reels. This action lifts the reels off the bottom cover and unlocks them for use.

- *Tape viewing windows*—Transparent windows in the top cover allow the user to see how much tape is in the cassette and where it is positioned on the reels.

- *Write-protection*—An external tab on the side of the cassette allows the cassette to be locked from recording. This prevents any undesired over-recording of important material.

All cassettes provide these features and sometimes more. Some systems are designed to accommodate different-sized cassettes, which requires changing the reel-hub spacing of the deck as the cassette is loaded. The deck does this by sensing the cassette size as it is inserted. As can be seen, a cassette is a fairly complex device; it is a marvel of mass production that they can be manufactured for only a few dollars.

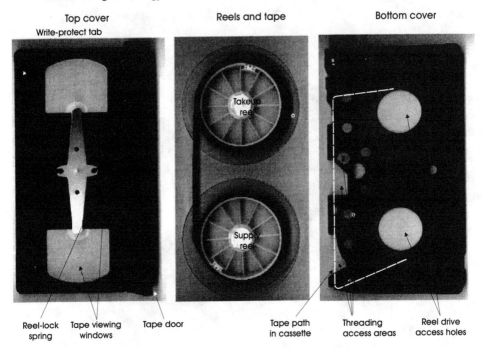

| Top cover | Reels and tape | Bottom cover |

Write-protect tab

Takeup reel

Supply reel

| Reel-lock spring | Tape viewing windows | Tape door | | Tape path in cassette | Threading access areas | Reel drive access holes |

Figure 5.13 A VHS cassette disassembled.

5.7 CASSETTE LOADING MECHANISMS

There are three basic steps to loading a cassette into a deck for use—a process sometimes called *automatic threading* (a name taken from the way it used to be done manually, but really referring only to the third step of loading). They are described in the sections below.

5.7.1 Cassette Insertion

Most stationary decks have a slot through which the cassette is inserted. Portable decks usually are loaded by opening a door and inserting the cassette into the door; this approach allows the next step to be accomplished manually through the action of closing the loaded door. In either case, the action of cassette insertion causes the tape door on the cassette to be opened.

5.7.2 Mechanism Engagement

This involves the engagement of the reel drives and the insertion of tape threading pins or guides into the cassette. In a slot-loading deck, it is done by lowering the cassette onto the deck elements after it has been inserted into the slot. The mechanism for this is usually called the *elevator*. In a door-loading deck, closing the door manually performs the elevator step.

5.7.3 Tape Threading

This is the final and most complex step of cassette loading. The tape is drawn from the cassette and placed into the deck's tape path, which includes wrapping it around the helical scanner.

In Figure 5.14(a), multiple pins and guides enter the cassette behind the tape during the engagement step; in threading, they move out of the cassette along paths that bring the tape into the desired path. It is usually necessary to have two guide assemblies that move out around the scanner, and the capstan pinch roller moving separately to its disengaged position near the capstan. Further moving guides may be necessary to achieve the tape wrap for the tension arms. Although this is an effective approach, it has many individually moving parts and is expensive.

The approach used in most home decks, shown in Figure 5.14(b), basically has a single moving part—a ring that rotates around the scanner, carrying all the necessary tape guides and rollers. In practice, there are additional movements, but they are all accomplished by cam action driven from the rotating ring. This design gives a very compact deck that is suitable for use in camcorders and stationary decks.

5.8 CONCLUSION

This chapter gave an overview of tape deck components and their application in various types of decks. Many other variations exist beyond those shown here. Chapter 10 gives more information about the tape decks of specific video recording systems.

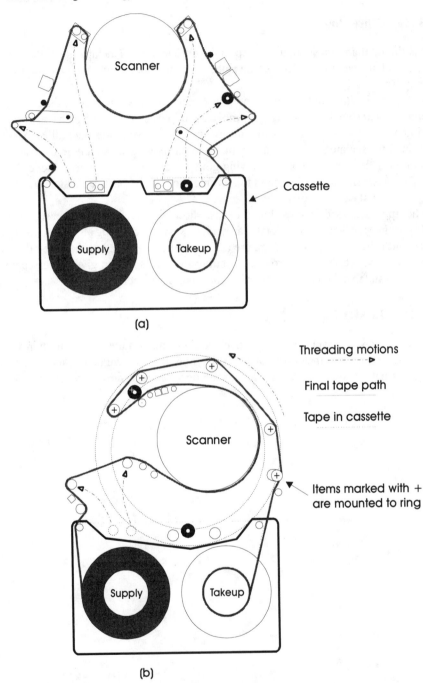

Figure 5.14 *Two approaches to threading tape from a cassette: (a) individually moving pins and (b) rotating ring.*

6

Disk Drives

In a disk-based recording system, the main mechanical assembly is the *disk drive*. This term is often used to refer to an entire disk recording system, but in this chapter, it means only the mechanical assembly and associated servomechanisms. Disk signal system electronics are covered in Chapters 7, 8, and 9.

The tasks of a disk drive are simple: it provides means for rotating one or more disks and positions one or more head assemblies to read or write on the disk surface. This chapter will describe how those tasks are accomplished for both magnetic and optical disk recorders or players.

Unlike a tape deck, which accesses a large area of recording medium by unrolling a tape, a disk drive is limited to the recording area on its one disk or, at most, the few disks that are mounted on the spindle. In video recording terms, this limited area of medium means that playing time may also be limited, or that there may be a trade-off between playing time and picture and sound quality. With the exception of the laser video disk, all video disk recorders are digital, which allows the trade-off mentioned above to be fully exploited by advanced signal processing (described in Section 7.3.4).

This chapter covers disk drive hardware, servo systems, and some of the performance considerations for video recording on disks.

6.1 THE TYPICAL DISK DRIVE

Figure 6.1 shows a typical disk drive with its principal components. Although the figure is typical for a magnetic hard drive, magnetic floppy drives, and optical drives share the same components.

- *Baseplate*—This supports all the parts of the disk drive.
- *Platter*—This is the recording medium. There may be one or more disks mounted permanently on the same spindle in a hard disk drive, but with floppy disks or optical disks, there is only one disk, which is usually removable.

119

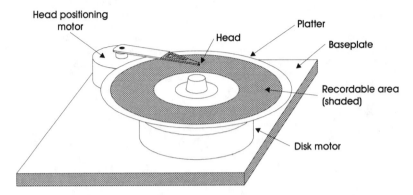

Figure 6.1 *Typical disk drive.*

- *Disk motor*—This rotates the disk and usually has servo control.
- *Head positioning motor*—As shown in the figure, a portion of the disk's surface is usable for recording. Spiral or concentric tracks are arranged to fill this area, and the head positioning motor or mechanism is capable of placing the head anywhere in the recordable area.
- *Head*—This is the record or replay head.

6.2 MAGNETIC DISK DRIVES

There are two types of magnetic disks based on the substrate material of the platter: flexible-substrate platters are called *floppy disks*, and rigid platters are called *hard disks*. Floppy disks are generally removable from the drive and are used for off-line storage in computers or other systems. Hard disk platters are usually captive in their drives, which are tied to a particular system for on-line storage. Removable hard disks (e.g., the Iomega 1 GB and 2 GB Jaz systems) are also available and are becoming popular in some segments of the PC market. Removable hard disks are tricky because state-of-the-art hard disk technology preferably requires operation in a sealed environment, which is awkward to accomplish with removable platters.

Another removable hard disk method is to remove an entire drive to change disks in a system.

6.2.1 Floppy Disk Drives

Floppy disk drives are currently not very suitable for real-time video recording because their data rates are too slow. However, that may not be the case in the future, so they are covered here.

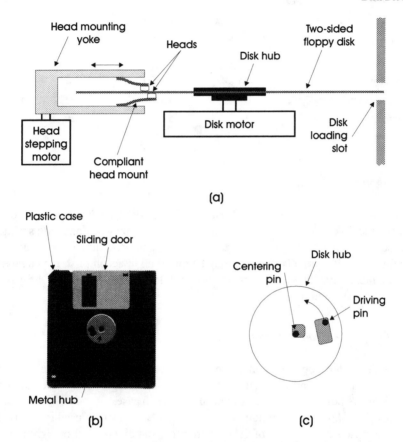

Figure 6.2 *A typical floppy disk drive: (a) cross-section of a typical drive, (b) the 3.5-in floppy diskette, and (c) centering of the 3.5-in diskette.*

Floppy disks today all have a hard plastic protective cover with a moving door for head access to the disk. The cover and door protect the disk from handling and seals the disk surface from dirt and dust under normal use conditions.

Figure 6.2(a) is a cross-sectional drawing of a typical floppy disk drive and Table 6.1 lists the performance of current systems. Figure 6.2(b) shows the 3.5-in floppy diskette, which is the most common diskette used in PCs. It has a metal hub, which is automatically centered by two pins in the drive, as shown in Figure 6.3(c). One pin enters the center hole to provide centering and the second pin enters the rectangular hole off-center in the hub. When the diskette is rotated by the second pin in the direction shown, the pin in the rectangular hole drives the center pin to the corner of the center hole, providing exact centering regardless of diskette and drive tolerances.

Table 6.1
Magnetic Floppy Disk Performance

System	Formatted capacity (MB)	Data rate (Mbps)	Seek time (ms)
3.5-in computer standard	1.44	0.5	93
Iomega Zip drive*	100	5	29

* Internal interface

6.2.1.1 Floppy Disk Loading

Disks are placed into the drive through a slot. After insertion, the disk's protective door is opened and the disk hub engages with the disk motor. This may be done with springs that are released when the disk is fully inserted; they move the disk down a short distance to engage the hub. Alternatively, the hub may be raised up to accomplish the engagement. Either way, most drives have an external button that the user pushes to eject the disk when it is no longer needed in the drive.

6.2.1.2 Floppy Disk Head-to-Disk Contact

The flexibility of the floppy disk medium is used in all systems to provide a controlled head-to-disk contact situation. Systems may have either single- or double-sided disks; Figure 6.3 shows how head contact is controlled in each case.

Figure 6.3(a) shows the common two-sided 1.4-MB floppy disk used in nearly all PC systems. The heads on each side of the disk are offset radially by four or eight tracks and they are mounted on ceramic sliders that are loaded onto the disk surfaces by a compliant suspension such that they push against each other to develop the head-contact force. The disk compliance allows the disk to move up or down to take up the mechanical tolerances of this. The wear resulting from good head contact limits the rotational speed of such a disk drive to about 300 rpm. Most disk drives are designed to unload the heads from the disk when not actually recording or playing, or, in low cost drives, the disk motor is simply stopped when not in use. This wear situation, of course, puts a limit on data capacity and data rate from the drive.

Figure 6.3(b) shows a single-sided disk configuration. Early systems used a pressure pad agains the disk surface opposite the head to provide the contact force but that was later eliminated as the industry went to two-sided disks.

Single-sided disks, however, are still used in high-density drives; the air flow around a disk rotating at high speed in close proximity to a polished metal plate supports the flexible disk and stabilizes it against vibration. A head may contact the disk surface by pressing against the stabilized disk, deflecting the disk to create the necessary contact force. This is known as the *Bernoulli* method. It is capable of providing good contact with limited wear, and thus allows greater disk rotational speeds—up to 2,000 or 3,000 rpm.

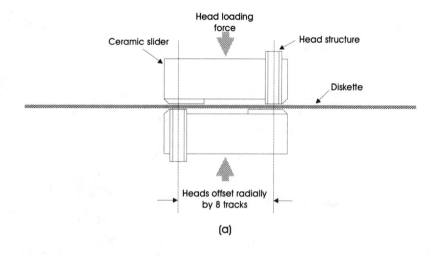

(a)

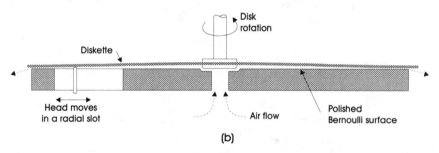

(b)

Figure 6.3 *Head to disk contact in floppy disks: (a) dual-head floppy disk and (d) Bernoulli plate*
stabilizer.

Another high-density head contact approach stabilizes the disk between two air films and
then has a "flying" head as used in hard disk drives (see Section 4.1.4).

6.2.1.3 Floppy Disk Head Positioning

Because of the magnetic azimuth issue, magnetic floppy disks generally use a linear motion of the head along a radius of the disk. A rotary-motion positioner has a changing
azimuth with head position, which is unacceptable in a system with interchangeable disks.
(Rotary positioners are used with hard disks, but there the heads are always using the
same captive disk and azimuth changes always cancel out.) Head positioning in most
floppy drives is purely mechanical with no servo action. This places a minimum on track
width, but this is consistent with low cost. In some current high-density floppy drives, a
head position servo is being used to allow smaller track pitch.

Many variations in linear head-positioning techniques have been developed by
striving for the best cost-performance results in floppy drives. These can be summarized

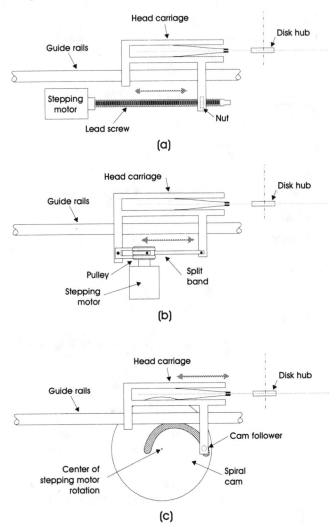

Figure 6.4 *Floppy-disk head positioning approaches: (a) lead screw, (b) split-band capstan, and (c) cam follower.*

in terms of three general approaches, shown in Figure 6.4. All head positioners have a track or rods that guide a head carrier as it moves across the disk surface. In the case of two-sided disks, the head carrier must move a head on each side of the disk. The variation is mostly in how the head carrier is moved along the track. The motor in each example is shown as a rotary stepping motor, which is most popular because of its low cost, but linear motors and voice-coil actuators have also been used. The three approaches are as follows:

1. *Lead screw*—Figure 6.4(a) shows a lead screw driven by a rotary stepping motor.

The nut on the lead screw directly moves the head carrier.

2. *Band-capstan*—This approach (Figure 6.4(b)), uses a metal band that wraps around a capstan driven by the stepping motor. By splitting the band and fixing it to the capstan, a very precise and repeatable drive is obtained.

3. *Cam*—In this approach, shown in Figure 6.4(c), the stepping motor drives a cam wheel that has a spiral slot holding a cam follower that moves the head carrier along its track as the cam wheel turns.

The band-capstan approach is most popular because it provides the best cost-performance factor. The other approaches are used in special cases where higher speed or stability is desired.

6.2.1.4 High-Density Head Positioning

For high density floppy disk drives for capacities of 100 MB or more, servo head positioning is required. The greatest problems arise from anisotropic expansion of the plastic disk substrate with temperature, which creates oval tracks, and disk mounting errors that cause eccentricity of the tracks. A track-following servo is needed when these inherent errors begin approaching the track width. These problems are similar in hard disk drives (see Section 6.2.2.1), except that they become important at lower track densities in floppy disk drives. However, the same servo approaches can be used.

Disk tracking servos all require the provision of additional information on the disk for use by the track-following detector. One method involves prerecording low-frequency information on the disk (which could be done at the factory if users never bulk erased their disks); for example, recording two signals at different frequencies side-by-side can provide a marked path that a tracking head follows by adjusting track position so that the output of the two frequencies is equal. This would indicate that the head is following over the junction between the two low-frequency tracks. A high-frequency data recording could then be placed there without any signal penetrating deep enough into the medium to damage the tracking signals. This takes only one head for both tracking and reading data. However, it requires a thick magnetic coating to record the low-frequency tracking signals, which is not suitable for the best recording of the high-frequency data signal.

If the head could be equipped with a fast-response actuator, such as the piezoelectric bimorph described in Section 5.5.5 for helical-scan recorders, the same dither-based automatic tracking scheme described there might be used to follow disk tracks. This has not been done for disk drives, probably because the necessary fast response would be difficult to obtain on a flying head, but a widely-used approach that provides servo control without having to dither the head records special blocks of track data (sectors) at different track positions. This is described in Section 6.2.2.3 for hard disk drives.

6.2.2 Hard Disk Drives

Figure 6.1 showed the components of a typical hard disk drive. With the rigid platter rotating at high speed, the flying-head technology described in Section 4.1.4 allows reliable noncontact operation at high recording densities. Operating in a closed, controlled environment, platter and head life are extremely long and recording densities continue to be increased with the introduction of new products.

6.2.2.1 Hard Disk Platters and Motors

Platters for hard disks were described in Section 3.3.2. Most high-capacity drives have multiple platters mounted on the same spindle. The spindle is driven at constant speed by a brushless DC motor having a phase-lock servo. Both sides of each platter are used for recording and there is usually one flying head per surface. All heads are moved by the same actuator and record concentric tracks at the same radial positions on each surface. This gives rise to the concept of a *cylinder* of data—referring to the three-dimensional appearance of the group of data tracks that are recorded at one setting of the head-positioning actuator.

The flying head technique only "flies" while the disk is rotating at normal speed. When the disk is stopped, the head must "land" somewhere on the disk surface. An area is generally provided for this at the inside or outside of the disk radius.

6.2.2.2 Hard Disk Head Positioning

Early hard disk drives used fixed head positioning to record or read each track. This meant that the unavoidable errors of head positioning, disk runout, vibrations, and so on had to be accommodated within a small fraction of the track width to avoid excessive mistracking loss of signal. However, this simple approach is impractical with the track densities used in modern hard disk drives; a head-positioning servo is mandatory. Many approaches have been used, but the most common servo used today is the *sector servo*, described below.

6.2.2.3 Sector Servo Head Positioning

Because of the need for random access to any section of the data, it is customary in both hard and floppy disk drives to divide the data into blocks, called *sectors*. All sectors generally contain a fixed amount of data but, because of the changing radius of different tracks, they are not all the same physical size on the disk. Since a fixed data rate cannot exceed the value supported by the smallest sectors (inner tracks), data are recorded at a lower density in the sectors of the outer tracks. Some drives partially compensate for this by increasing the data rate as the track radius increases, which means that outer tracks have more sectors than inner tracks. Because there should be an integral number of sectors per track, the data rate is changed in steps for groups of tracks; still, the data capacity can be doubled from that available with a fixed data rate. This type of sector layout is shown in Figure 6.5(a), which shows four sector groups with 12, 15, 20, and 24 sectors per track (these are low numbers, chosen to simplify the figure).

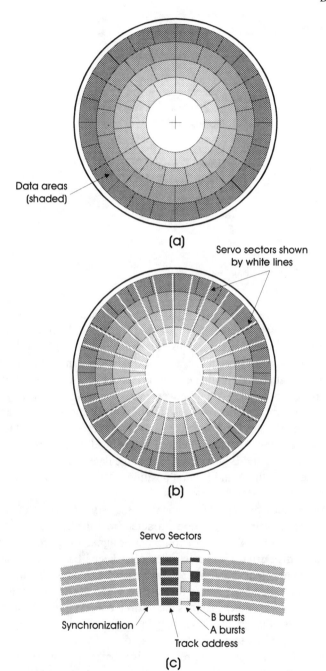

Figure 6.5 *Principle of head positioning with a sector servo: (a) data sectors in four groups, (b) servo sectors shown in white, and (c) detail of servo sectors.*

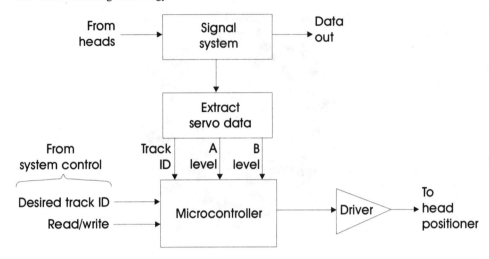

Figure 6.6 *Block diagram of a sector servo.*

To add servo information to this, small sectors are placed at fixed radial positions around the disk as shown in Figure 6.5(b). These sectors are recorded with information that allows them to be identified as the disk is being read, as well as information that is positioned halfway between the final track locations. This information is prerecorded when the disk is formatted and should never be affected by the data recording because that is always controlled to use the space between servo sectors. As can be seen in Figure 6.5(c), the servo sectors include a synchronization block that can be read at any head position, a track identification code, and the A and B bursts that are offset to either side of the desired track locations. As with optical disks, the head position is servo controlled to make the output of the A and B burst equal. Thus, the data track is recorded along a path that is centered between the A and B burst positions. This is done in both read and write modes.

Figure 6.6 is a block diagram of a typical sector servo. It uses a microcontroller to perform all processing, using information extracted from the head data in the signal system. (A microcontroller is a small microcomputer designed to be embedded into another product, such as a disk drive.) The microcontroller keeps the currently accessed track number, so it knows how to respond when a track change is requested. The head positioner is moved to the approximate location of the new track and the data from the servo sectors is examined. If the track number matches the desired one, the servo mode is activated to match the amplitudes of the A and B burst signals.

The sampling rate of the sector servo is equal to the disk rotation rate times the number of servo sectors on the disk. Faster response is possible with more servo sectors, but, of course, each servo sector subtracts from the disk area available for data recording, so a compromise has to be made. A typical number of servo sectors is 64.

Table 6.2
Storage Requirements of Different Video Formats

Video format	Resolution (pixels)	Frame rate Hz/I or P**	Data rate (Mbps)	min/GB	GB/hr
Compressed					
MPEG-1	320 × 240	30 P	1.2	14	4.3
MPEG-2 SDTV	640 × 480	60 I	4	4.3	14
MPEG-2 HDTV	1920 × 1080 I	60 I	20	0.85	71
Uncompressed					
Digitized NTSC- 4×fSC		60	92*†	0.18	330
Rec. 601-4 4:2:2	720 × 480	60 I	216†	0.08	750

*   Blanking intervals not transmitted.
** I—interlaced scan, P—progressive scan.
†   A single hard disk drive cannot achieve this rate.

6.2.2.4 Hard Disks for Video Recording

Nonremovable hard disks are suitable for video recording only in applications where the recording is kept within a single system and media does not have to be exchanged between systems, unless whole drives are removed. There are many important applications where this limitation can be accepted; generally these are systems where video is delivered to external lines on demand, but the record medium itself never has to be taken out of the system. These include video and audio editing, video servers, broadcast playing-to-air, certain video camera applications, and others. The video recording time requirement varies among these applications and it is useful to examine the trade-off of storage requirements versus picture quality for different video standards. This is shown in Table 6.2. Removable hard disks have smaller capacities than nonremovables. They are suited for recording video within their capacity limits.

Considering that the small, inexpensive hard disk drives now available for personal computers go up to 12 GB capacity or more, it can be seen that these drives are useful for recording significant amounts of video when using the MPEG-1 or MPEG-2 compression standards. However, they are not up to handling any significant amount of HDTV video, even when compressed. Storing uncompressed video on these small drives is out of the question.

Some of the applications for hard disk storage of video can afford to use more expensive technology for their storage; these use multiple hard disk drives (called *arrays*) to obtain both increased capacity and higher data rates—all the high-end video editing systems use disk arrays. Considering the need for random access in editing (see Chapter 8), there is no other practical recording solution at any price. The cost of editing systems is driven heavily by video quality specifications.

6.3 OPTICAL DISK DRIVES

An optical disk drive contains the same component types as a magnetic drive: platter, disk motor, head, and head positioner. Essentially all optical disk drives use removable disks, so the designs contain many of the same features as magnetic floppy disk drives. It is not necessary to repeat here the description of those components that have been covered in this chapter and elsewhere, but some special considerations will be discussed.

6.3.1 Disk Motor and Servo Issues

Optical disk drives that use CLV or have a CLV mode have the additional requirement that disk motor speed must be varied with head position. This is true for all the CD-based drives. The use of variable speed, however, has been exploited in CD-ROM drives to increase their access times and data rate capability far beyond that of the original CD specification. It is interesting to examine how this is done.

The CD-digital audio specification (CD-DA) called for CLV operation at 1.3 m/s, which must be maintained accurately in audio players to keep data flowing smoothly. The disc rotational speed must therefore change from 530 rpm at the inner tracks to 200 rpm at the outer tracks. Early CD-ROM drives operated the same way. When used for general data recording, however, there is no requirement on a CD-ROM drive to maintain this speed—in fact, it is better to increase the rotational speed because that improves access time and moves the data faster into the host computer. Modern CD-ROM drives are classified according to how much faster they are than the original CD-DA: 2×, 4×, 8×, and so on. The current state of the art is 32×, although that is achieved only at the outer tracks; inner tracks are slower because the disc rotation cannot be speeded up enough to hold 32× there. 32× on the outer tracks requires disc rotation of 6,400 rpm.

Thus, the high-speed CD-ROM drives no longer give a priority to CLV operation. Instead, they use different algorithms to optimize both data rate and access time and the actual linear track speed is not important [1].

Video generally requires a fairly constant data rate; with a very fast CD-ROM, data will be delivered in bursts and the host system must provide buffering to hold the bursts of data for playback at the specified rate. This allows the CD-ROM drive to wait until the buffer is getting low before delivering another burst of data to refill the buffer. Depending on the rate of the video, and the rate of the CD-ROM and its access time, this may require 1 or more MB of RAM for the buffer, which is not very expensive these days and may be included in the CD-ROM drive.

6.3.2 Optical Head Positioning

The most common optical head positioning servo was described in Section 2.3.1.1. By splitting the optical beam into three spots with a diffraction grating, the track is located by making the two side spots read at the edges of the recorded track; correct tracking is indicated when the outputs from the two side spots become equal.

The optical read or recording head is a large assembly, as was shown in Figure 4.10. This assembly must be moved from track to track by the head positioning motor. The typical approach is similar to that of the magnetic floppy disk drive: the head assembly moves on guide rails and the motor is coupled to move it along the rails. This, however, would be much too slow for a tracking servo, which has to respond to variations occurring in a small fraction of one disc rotation. Because the fast tracking actuator is on the objective lens, however, the slow response of moving the entire head assembly is not a problem. The head movement is used to change tracks and to keep the lens tracking actuator within its range.

6.3.3 Optical Disk Changers

In the CD-DA market, it was attractive to offer multidisc players that would allow users to play hours of music without attention to the CD player. Players holding 3, 5, or even hundreds of audio CDs are available. Similar mechanisms have been made in CD-ROM format but are not very popular, probably because accessing from one disc to another is quite slow. This takes away much of the convenience of having multiple discs on line. CD drives are so inexpensive now that it is also feasible to have multiple drives on line in some applications.

The same kinds of mechanisms may prove to be popular in home video players, although it seems unlikely that viewers will find need for such long playing sessions. Since home video is going to be DVD, which can play 8 or more hours of video from one disc, there may not be much need for a DVD multidisc player.

6.4 CONCLUSION

Tape decks and disk drives are the principal mechanisms of video recording. They have been developed to a high degree of sophistication and cost effectiveness for high-volume markets such as the home and PCs. Those developments have spun off into professional systems, which use the same mechanisms and benefit from the massive investment in design and tooling for the volume markets. Professional systems, however, generally include much greater electronic sophistication to achieve their more demanding performance and feature goals.

REFERENCE

[1] Stan, S. G., and Bakx, J. L., "Adaptive-Speed Algorithms for CD-ROM Systems," *IEEE Transactions on Consumer Electronics*, Vol. 42, No. 1, Feb. 1996.

7

Recorder Signal Processing

Video recorders would not exist without signal processing techniques; these provide not only basic record and replay functions, but they can detect and correct distortions or errors, compress and decompress signals to improve recording efficiency, or add special record and replay features.

Recording processes are not a good match to video signals, but many techniques are available for solving that problem in both analog and digital systems. This chapter discusses those techniques.

7.1 INTRODUCTION

Video signal processing differs greatly between analog and digital systems, as shown by the block diagrams in Figure 7.1. Although digital processing appears much more complex than analog processing, it is not necessarily more costly because it may be accomplished with just a few *application-specific integrated circuits* (ASICs).

7.1.1 Analog Signal Processing

The typical recording channel is highly nonlinear and does not have good low frequency response. In analog video recording systems, frequency modulation (FM) is generally used to accommodate the channel properties (see Section 2.2.6.2).

On replay, as shown in Figure 7.1(a), it is necessary to carefully equalize the channel frequency response before the FM demodulator. After demodulation, most systems have a *dropout compensator* (DOC) to minimize the effect of the momentary signal loss caused by occasional physical defects in the record medium, called *dropouts*. The time stability of an analog recorder depends directly on the mechanical motion of the scanning mechanism. Home recorders are designed to provide mechanical motion that does not change too rapidly, allowing the synchronization circuits of a typical TV receiver to follow the

133

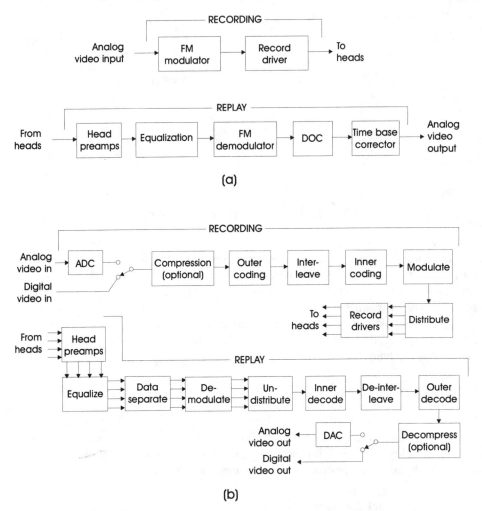

Figure 7.1 *Typical (a) analog and (b) digital recorder signal processing block diagrams.*

time variations. Professional recorders, however, generally have a *time base compensator* (TBC) to correct signal time variations so that recorder output signals are precisely synchronized with other signals in the system. This also prevents accumulation of time base errors in multiple generations.

7.1.2 Digital Signal Processing

The major advantage of digital recording is that signal processing can be done to make the recorder performance almost totally independent of the signal channel properties. Typical digital recorder processing is shown in the block diagram of Figure 7.1(b).

Many digital recorders can accept either analog or digital signals. For analog inputs, an ADC converts the signal to a digital format; the rest of the recorder processing is all digital, except that a DAC may be provided for analog outputs. The steps of digital processing are as follows (some systems may perform these steps in different order):

- Video compression is an important feature of digital video systems; it is often necessary to reduce the data rate for specific requirements. However, compression is not desirable in recording systems that will be used for multiple generations because compression artifacts may accumulate as generations are added. Thus, professional systems tend to not use compression or only use it a little, while home systems almost always use it.

- Randomizing is used to reduce or eliminate the dc content of digitized video.

- Another advantage of digital systems is the capability for error protection, which can totally eliminate the effects of channel noise within the design range of the system. In recorders, this is generally done twice, with one error coding within the other error coding. This gives rise to the terminology of *inner* and *outer* coding.

- Interleaving is a process of rearranging the data sequence to break up bursts of channel errors into separate error correction blocks. It is done between the inner and outer error coding.

- Data distribution is used in recorders that have more than one physical recording channel. It generally uses a strategy that allows errors in one channel to be corrected or compensated by data taken from another channel.

In replay, the recovered signal is amplified and equalized, and data separation recovers the recorded bit stream from the channel signal. The other steps of processing that were done during recording are undone in reverse order to recover the original signal with error correction.

The rest of this chapter discusses the signal processing steps in more detail.

7.2 PREAMPS AND DRIVERS

These devices interface between recording or replay heads and the rest of the system.

7.2.1 Objectives of Preamps and Drivers

The objectives of preamps and drivers include the following:

- *Convert between the head and the processing system operating levels.* During recording, an amplifier is used to increase the power level to that required by the head. In replay, the preamp is an amplifier to bring the head level up to the system's operating level. In this case, noise is an important consideration because head output levels are usually low and amplifier noise should be minimized.

- *Provide the necessary bandwidth for the system operation.* In a record driver, the frequency response is usually made flat over the desired bandwidth, but in replay the preamp may have a characteristic response that will have to be equalized later in the system.

- *Provide suitable linearity so that the desired signals are sent to or received from the head without distortion.* Although most recording channels themselves are not very linear, it is desirable for the amplifiers involved to be linear so that they are not sources of additional distortions that might create spurious frequency components added to the signal. Any nonlinearity in the system (such as that caused by the recording mechanism itself or by FM limiters, etc.) must be carefully specified and controlled to avoid undesired effects. This is true for both analog and digital systems.

7.2.2 Magnetic Head Preamps

The output from magnetic heads is at a low enough level that preamp design must strive for low noise from the amplifier. Ordinarily, a magnetic recording system should be limited by noise from the medium rather than the amplifier, but the amplifier noise can become a factor if the amplifier is not well designed.

Head preamps for helical scanners are usually placed on the drum and rotate with the heads. That way, the lead length from the head to the preamp can be very short, which is best for low noise and avoiding stray pickup. Also, a preamp on the drum raises the operating level for the rotary transformers that couple the rotating heads to the rest of the system, eliminating the rotary transformers as a source of noise.

In a magnetic head preamp, the head's inductance will resonate with its stray capacitance. If the resonance is within the signal band, it must be equalized in the preamp or later in the system. However, this is a critical adjustment that may have to be changed as heads wear; modern magnetic head preamps are usually designed with a low input impedance that reduces the effect of head resonance and the need for critical adjustment.

7.2.3 Optical Preamps

The signal levels from the optical photodetectors are high enough that preamp noise is not usually a factor. It is still good practice to place the preamp first stage on the optical head assembly to minimize lead lengths and avoid pickup of interference.

7.2.4 Magnetic Record Drivers

When recording, it is necessary to drive the heads strongly enough so that the recording is "optimized," meaning that the record current is adjusted to give the maximum signal output on playback. In analog recorders, this is determined by observing the FM carrier output; in digital systems it is done by adjusting for the best eye pattern (see Section 7.3.5). The optimization effect occurs because the recording of short wavelengths will actually fall off at higher levels because of saturation in heads or medium. This should not be limited by the record driver.

Delivering high currents at high frequencies can cause a heat problem in record drivers. Since the signal levels are high, lead lengths are not an important issue, and the record drivers can be located where they can be properly cooled and where their heat will not affect other temperature-sensitive parts of the system.

7.3 MODULATION AND DEMODULATION

Video recorders generally use modulation to make the signal suitable for the recording channel. An exception is in analog audio magnetic recorders where bias recording is commonly used to handle the audio signal directly. However, as explained in Section 2.2.6.1, this also can be viewed as a form of modulation that takes place at the head-tape interface.

7.3.1 Reasons for Using Modulation

Modulation is used for the following reasons:

- To match the frequency range of the signal to the recording system. For example, most recording systems do not have good low-frequency response and have zero response at dc. Modulation, such as analog FM, translates the signal frequency range upward so low-frequency response is not required in the recorder. This also has the effect of reducing the ratio between the highest and lowest frequencies in the band, which simplifies channel equalization for frequency response.
- To make the system response immune to the nonlinearity of the recording process. This is true for both analog and digital systems. For example, with analog FM, the system linearity depends only on the linearity of the frequency modulator and demodulator. In a digital system, system linearity depends only on the linearity of the ADC and DAC.
- In the case of digital systems, the modulation provides for clock extraction on replay so that data separation can work correctly.

Especially in digital systems, it is a little difficult to decide where "modulation" begins and ends in the signal chain because there are so many steps. In this book, digital modulation is taken to mean only those processing steps that have the above objectives. Thus, it does not include processes such as randomizing, error protection, or interleaving. In recording, the modulation is usually the last step of processing before data distribution and record drivers, whereas in replay, demodulation is the first step after data separation.

7.3.2 Analog Frequency Modulation

FM was described in Section 2.2.6.2. This section describes some additional considerations about the use of FM.

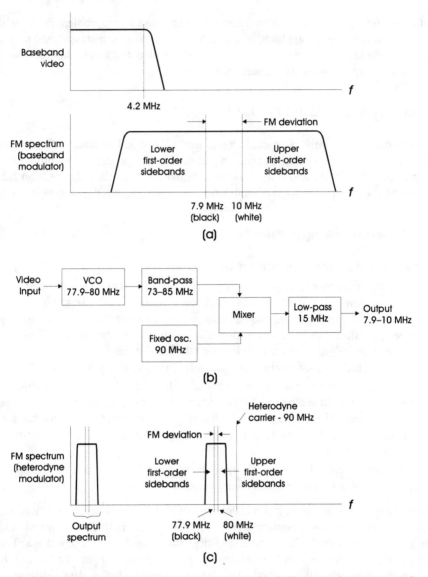

Figure 7.2 *FM modulators: (a) baseband modulation and (b) block diagram and (c) spectrum of a heterodyne modulator.*

7.3.2.1 FM Modulator Design

The frequency spectrum of a frequency-modulated signal contains multiple sidebands on each side of the carrier. However, the limited bandwidth of recording systems can carry only the first-order sidebands of the highest modulating frequency. Thus, an objective of an FM modulator for recording is to assure that higher-order sidebands do not get folded back through zero frequency into the recorded pass band, where they would cause spurious patterns in the recovered signal. This can be a problem if the modulation is accomplished by changing the frequency of an oscillator running directly at the frequency to be recorded, as shown in Figure 7.2(a).

The problem is overcome by modulating an oscillator at a higher carrier frequency and heterodyning the signal down to the desired frequency range after filtering to remove unwanted sidebands, as shown in the block diagram of Figure 7.2(b) and the spectrum diagram of Figure 7.2(c). A voltage-controlled oscillator (VCO) operates in the 77.9–80 MHz range and is frequency-modulated by the video signal. The output of this oscillator passes through a band-pass filter to remove frequency components outside the range that can be recorded. Then, the signal is mixed with a fixed oscillator and the difference component is selected by a low-pass filter.

The specifications for frequency modulation by a video signal require that the video black level be maintained at a fixed frequency under all signal conditions. This is accomplished by sampling the output frequency of the modulator during the blanking back porch interval and comparing it to a fixed reference to develop a feedback signal that adjusts the dc component of the input video signal.

7.3.2.2 Modulator Testing

The processes of modulation and demodulation in cascade should be as transparent as possible to the video signal. Most systems have a mode to directly connect modulator and demodulator together (bypassing recording) to test their performance. This is known as electronics-to-electronics mode (E-E). It is often implemented to occur automatically when the recorder is in STOP mode and nothing is being recorded or played. Recorders that cannot play back while recording also implement E-E mode during recording. Thus, a monitor connected to the output of the recorder will show the incoming signal during recording.

7.3.3 Color-Under Modulation

High-performance analog FM recording requires a channel bandwidth of somewhat more than three times the video bandwidth plus the FM black-to-white deviation range. Since recorder bandwidth is expensive, home video recorder designers look for ways to reduce the recorder bandwidth and still have acceptable video performance for the home market. The most popular solution is the *color-under* system, where the chrominance and luminance signals are separated and recorded in two frequency bands in the same channel. This is shown in Figure 7.3.

The chrominance components are separated from the luminance components and downconverted by heterodyning with a fixed oscillator at approximately 4.2 MHz to a carrier frequency of typically 562.5 kHz. The bandwidth of this is limited to about 1 MHz. The luminance components are reduced to a bandwidth of about 2.5 MHz and modulate an FM carrier at around 5 MHz. The two signals are combined for recording. The higher-

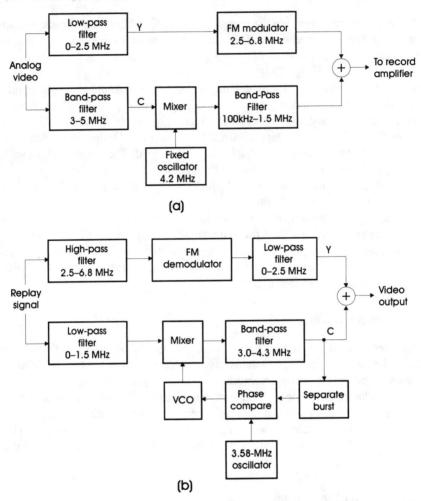

Figure 7.3 *Block diagrams of the color-under system: (a) recording and (b) replay.*

frequency FM component acts as a bias signal for linearly recording the lower-frequency color component.

On replay, the FM luminance signal and the color components are filter-separated. The luminance signal is demodulated and the color components are up-converted to the correct color subcarrier frequency, and the result is combined with the demodulated luminance. The up-converting oscillator frequency is derived from the tape signals by separating the color burst and comparing it to a fixed oscillator, which then controls the up-converting frequency. This reduces the time base errors in the chrominance signal.

Notice that using the color-under technique with azimuth recording (see Section 3.2.1.1) requires special consideration because the down-converted color component signal con-

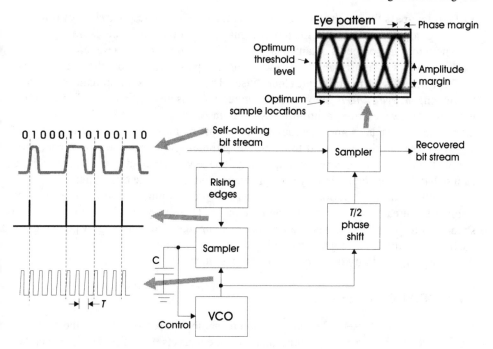

Figure 7.4 *Clock recovery and data separation.*

tains long-wavelength components where the azimuth technique is not effective in reducing crosstalk between channels (see Section 10.1.2.3).

7.3.4 Digital Modulation

Methods for digital modulation, sometimes called *encoding*, were discussed in Section 2.2.7.2. A good modulation method is capable of receiving an arbitrary bit stream and making it acceptable for efficient recording and providing for clock recovery on replay. The format of the bit stream is determined by the processes that preceded modulation, such as video compression and error protection coding. Modulation does not care about the incoming data format; its objective is simply to provide reliable recovery of the data on replay, so that subsequent processing can recover the information content and error-correct it.

7.3.5 Clock Recovery and Data Separation

Video recorders always use a self-clocking modulation method; the bit clock on replay is recovered by a *phase-lock loop* (PLL) operating from the data itself, as shown in Figure 7.4. In the figure, the rising edges of the incoming bit stream sample a reference waveform from the VCO. This creates a phase lock between the edges in the bit stream and the

VCO. A low-pass filter in the loop causes the extracted clock to respond slowly to changes of timing, so jitter and noise in the replay signal are averaged in the clock timing.

The actual data output is obtained (separated) by sampling the data waveform at points where the levels are most stable. This is typically at a point shifted from the PLL sampling by $T/2$, where T is the period of the clock pulses. The effectiveness of this can be observed by viewing an *eye pattern*, shown in the figure, which is an oscilloscopic display of the incoming data waveform synchronized by the extracted clock signal. Amplitude and phase jitter make the display somewhat fuzzy, but there are regions where the display is free of signal—these are the eyes, and they are the optimum places for the data separator to sample the waveform. A very noisy system will have very small eyes in its pattern, indicating that data separation is very critical and may not be reliable over time.

Following data separation, the demodulation process must be done to recover the original bit stream as it came into the modulator. There may still be errors in this bit stream, corresponding to faults in the record medium, the record and replay processes, or the data separation. The objective of this part of the system is to keep those errors within the correcting capability of the error protection system.

7.4 VIDEO COMPRESSION

More than anything else, video compression technology is responsible for the spread of digital video recording and transmission into home markets. Digital video, however, has been embraced by the professional markets for a different reason—its high quality, reliable performance. To many professional users, that precludes the use of compression because compression implies some trade-offs compared to the highest possible picture quality. Many professional users will pay higher prices for uncompressed digital video to avoid the possibility of any degradation in processing or recording.

7.4.1 Basis for Video Compression

Scanned video sequences offer several opportunities for compression, falling into two categories: redundancy and nonvisibility. The goal of a compression system is to detect these two situations and remove them, thus reducing the size of the data stream.

Redundancy arises from the sequential nature of scanning and can be either spatial or temporal. Spatial redundancy occurs when the same information is transmitted by adjacent pixels or scanning lines, and temporal redundancy is when the same information is transmitted by successive frames.

Nonvisibility is a little harder to define. In the simplest case, it is picture detail that a typical viewer cannot see under the specified viewing conditions. For example, there is no need to transmit detail that a viewer will not see because of his or her acuity limits. A more complex case of nonvisibility occurs because a viewer will perceive fewer levels of amplitude or color gradation as the resolution approaches acuity limits. Compression systems exploit these properties of human vision.

A certain amount of compression can be achieved by *lossless* techniques, meaning that the decompressed data stream is exactly the same as the original data stream without compression. However, this is limited only to the removal of redundancy, and the degree of compression using this method is small. Far greater compression is achieved by *lossy* methods, where the decompressed signal is only an approximation of the original, compromised in ways that the viewer hopefully will not see. Lossy compression can exploit the properties of nonvisibility.

The successful removal of redundancy and nonvisibility depends on the specification of viewing conditions for the final picture. If a viewer sits closer to a display than specified for the system, which he or she is free to do, artifacts from the compression may become visible. Likewise, if video that has been compressed and decompressed is processed in ways that depend on normally nonvisible fine details, this may enhance the compression artifacts, which may then be seen by the viewer. This latter situation is the reason that professional video producers do not want compression used in their system—the processes of editing, special effects, or image enhancement can amplify the compression artifacts that might otherwise be invisible.

Compression artifacts also can accumulate with repeated compression-decompression cycles as may occur in multigeneration recording. Of course, this would not happen in multigenerations if the signals were not decompressed, but many editing and postproduction processes require operating on uncompressed signals, so the compression-decompression cycles cannot be avoided.

7.4.2 Degree of Compression

It is useful to specify the effectiveness of a compression system by talking about its *compression ratio*, which is the ratio between the average incoming data rate and the average outgoing data rate from a compressor. This can be misleading, however, if not done carefully because there may be a significant loss of picture quality when going through a lossy compression-decompression process. For example, one way of reducing data rate is simply to reduce the resolution by throwing away pixels and lines (subsampling). Assuming that the viewing conditions allowed the full resolution of the incoming signal to be seen, the loss in visible picture quality by subsampling is directly proportional to the reduction of data rate. This is not valid compression because no effort has been made to retain the original picture quality.

In spite of the difficulties, compression ratio is widely used, but for lossy systems, it must always be accompanied by an evaluation of the compressed picture quality.

7.4.3 General Video Compression Techniques

There are many general methods for compression, which are often combined in a system. These are covered in this section. Further information can be found in [1].

7.4.3.1 Lossless Techniques

Lossless compression is seldom used by itself for audio or video because its degree of compression is small, but it is often used as a part of a lossy compression system. Some of the techniques are as follows.

- *Run-length encoding (RLE)*—In image data, it is common for the same pixel value to be repeated in uniform areas of the image. This situation can be compressed by replacing a series of identical values by the value and a count of how many times to repeat it. To identify places where this has been done, an *escape code* is used, which is a specific pixel value designated for this purpose. Thus, a single RLE event requires three pixel spaces, so it is only effective to code runs that are more than three repeated values. Note that the escape code does not have to be reserved from appearing in the image data; when that happens, two escape codes are transmitted together to indicate that the data value equals the escape code.

- *Pattern matching*—The concept of RLE can be improved by coding *patterns* of data instead of just repeated values. A table of patterns and corresponding escape sequences is generated; in the outgoing data stream, the patterns are replaced by their escape sequences. Of course, the pattern table (dictionary) must also be transmitted. Pattern matching requires much more processing during compression than simple RLE.

- *Statistical coding*—It is common in image data for different data values to have different frequencies of occurrence. A coding system can be set up to transmit high-occurrence values with a short bit code and lesser-occurrence values with longer bit codes. The most common of these techniques is *Huffman* coding, which is widely used in video compression systems to provide additional compression on data that has already been processed by other methods.

Lossless approaches are also useful as lossy techniques. For example, a pattern-matching system may only look for approximate matches before assigning a matching code to a particular data pattern.

7.4.3.2 Lossy Techniques

Some of the techniques of lossy compression are as follows.

- *Truncation*—Video pixels usually are assigned 8 or more bits each. In some viewing situations, however, this amount of gray scale capability may not be visible. By reducing (truncating) the number of bits per pixel, some compression is possible.

- *Subsampling*—Because the eye's acuity for colors is less than its gray scale acuity, color difference signals do not need the same resolution as the luminance signal. This is acknowledged by sampling color-differences at a submultiple of the luminance sampling rate (see Section 1.6.2.1). Subsampling vertically can also be done by reducing the number of lines sampled for the color-difference signals.

- *Color tables*—When displaying still images, it may be acceptable to reduce the number of different colors needed to display the image. For example, many computer displays use only 256 different colors, which are identified by a *color table* containing the actual color values to be displayed for any of the values of an 8-bit pixel. This method is enhanced by selecting the 256 colors specifically for each image displayed and sometimes by the use of dither to combine colors of adjacent pixels. However, an 8-bit color table is not very applicable to motion video because the changing of colors in successive frames produces a very disturbing color noise effect. For motion video, it is better to not use color tables but, instead, rely on other compression methods.

- *Differential coding*—In both audio and video data, the amount of change in value from one sample to the next has a probability distribution that falls off with amplitude of change. Thus, small changes in value are more probable than large changes, and it may be advantageous to use *differential coding* to code the change in sample value rather than the values themselves. Since the changes are mostly small, they can be coded with fewer bits than the samples themselves, resulting in compression. This has a problem when the change is too large for the coding range, giving rise to an artifact called *slope overload.*

- *Predictive coding*—Sometimes it is possible to make a predictor circuit that tries to predict the next sample value by examining previous samples. Then, the real next value can be compared to the predictor output and only the difference is coded. At the decompressor, an identical predictor examines the flow of data and adds the coded difference to the predicted values. The success of this depends on the effectiveness of the predictor module, which must exploit the characteristics of the data stream.

- *Transform coding*—One of the most powerful compression techniques involves transforming the data to a different format where the redundancy that might be removed is better exposed. One of the most widely used transform coding methods is the *discrete cosine transform* (DCT), which takes a two-dimensional block of pixel values (usually 8 × 8) and transforms it to an equal-sized array of frequency component coefficients. This is shown in Figure 7.5.

The DCT has several important features. The pixel values, shown as gray scale in Figure 7.5(a), are transformed to an array of spatial frequency coefficients, shown in Figure 7.5(b). These are arranged in an array of ascending spatial frequency, and if read out in the zig-zag pattern shown, they typically create a data pattern of descending values, with many of the later values being zero. Since the higher spatial frequencies in the image are less visible to the viewer, their amplitudes can be represented more coarsely by quantizing them with fewer bits. When the values reach zero, an *end-of-block* (EOB) symbol can be transmitted to tell the decompressor that all remaining coefficients of the array are zero. When the decompressor applies the reverse DCT, the pixel values are reconstructed, but with the compromises represented by the coarse quantization. If the quantization

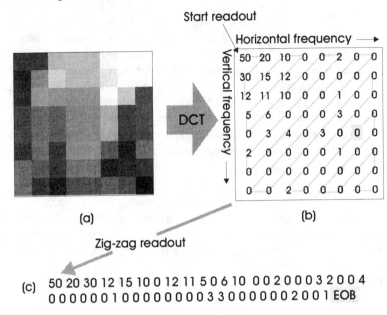

Figure 7.5 *The discrete cosine transform: 8×8 block of (a) pixels and (b) coefficients, and (c) corresponding data pattern. Reproduced with permission from [1].*

parameters are chosen carefully, the compromises in the reconstructed pixels will not be visible under normal viewing conditions.

- *Motion compensation*—All the techniques discussed so far deal only with compression of one picture or video frame at a time. *Motion compensation* is a technique for compressing the data temporally by exploiting the redundancy from frame to frame of the video stream. The basic idea is to divide each frame into blocks of pixels and then examine the previous frame to find whether any of the blocks are already present. If so, the block does not have to be retransmitted; the block from the previous frame can be moved into position for the new frame. This involves a massive amount of analysis, which is actually the limiting factor in how effective motion compensation can be.

There are many variations in the motion compensation technique. Depending on what type of motion may be in the scene, blocks in a new frame may be simply moved from the previous frame, or they might be rotated, shrunk, or expanded. In any case, the amount of motion is limited, so the previous frame only needs to be searched in a small range around the corresponding position of a block in the new frame. This limit is called the *search range*; the larger it is, the faster the motion that may be compressed. Motion compensation can also be done *bidirectionally* (B), where a block is searched in both previous and next frames. This affects the order of transmission of frames since both previous and next frames must be available before bidirectional searching is possible. This is shown in Figure 7.6.

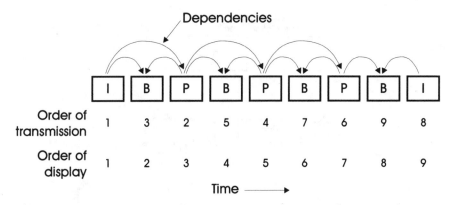

Figure 7.6 *Typical frame sequence for MPEG compression. Reproduced with permission from [1].*

Another caveat of motion compensation is that the frame-to-frame dependence can make it difficult to edit the data stream at any frame position. For this reason and also to prevent the propagation of a single frame error into multiple frames, it is normal practice to periodically insert *intracoded* (I) frames into the sequence. These are frames that do not depend on other frames, and they serve as benchmarks from which the motion compensation process can proceed without additional dependencies.

There are other compression techniques, but most of them are combinations or variations of these basic methods.

7.4.4 Video Compression Algorithms

As seen in the previous section, many techniques are available for compression of video; most systems use a combination of several methods, which is referred to as an *algorithm*. There are infinite possibilities for different algorithms. Because a decompressor must know explicitly how the incoming signal was compressed, the matter of signal standards is very important in the compression field. Much industry effort has been devoted to this problem and standards are available that provide a wide range of possibilities. The most widely used standards are those in the MPEG series (see Section 1.6.2.6). These standards support several algorithms; the definition of the algorithm is contained in the data stream so that an MPEG decompressor can handle different algorithms in different situations.

7.4.4.1 JPEG and MPEG

Before discussing the MPEG algorithms, the *Joint Photographic Expert Group* (JPEG) still-picture compression standards should be mentioned. JPEG is a working group of the ISO/IEC; the standard takes their name. JPEG is often referred to as a "tool box" of compression techniques because it has modes and options that cover a wide range of gray scale and color image compression applications. It provides both lossless and lossy modes,

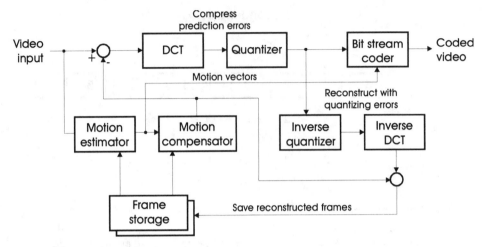

Figure 7.7 *MPEG compression processing. Reproduced with permission from [1].*

as well as modes where the image can be displayed progressively at higher resolution, speeding up initial access to an image. JPEG compresses each image separately using (in lossy mode) a DCT-based algorithm; it can be used for motion video by separately compressing each frame of a video sequence; this compares to the use of intracoded frames, mentioned above. The use of JPEG compression for a motion video sequence is called *motion-JPEG*.

MPEG uses DCT-based lossy compression techniques similar to JPEG, but it allows for motion compensation as well. The type of compression may be chosen on a frame-by-frame basis, mixing intracoded (I) and forward (P) or bidirectional (B) prediction motion compensation as desired by the application. One example of this is shown in Figure 7.6, showing the frame dependencies and transmission order effects for a sequence involving all three frame modes.

Figure 7.7 shows the basic processing of MPEG video compression. Incoming video frames, divided into blocks, are processed for motion compensation. Blocks that are predictable from already available frame data are transmitted as a difference signal obtained by comparing the block to the predicted information. Thus, the prediction does not have to be perfect to make a near-perfect reproduction because difference information is coded. Blocks that cannot be predicted are transmitted whole; both predicted and whole blocks pass through a DCT processor and quantizer. The output of the quantizer is then decoded by an inverse quantizer and an inverse DCT, and the result is combined with the motion compensator information to generate reconstructed frames, which are stored for use in compressing subsequent frames.

At this writing, two versions of the MPEG standard have been completed: MPEG-1 and MPEG-2. MPEG-1 was developed for compatibility with the early CD-ROM drives and provides a resolution of 320 × 240 pixels at a data rate of 1.2 Mbps; it gives results roughly comparable to home VCR performance. MPEG-2 came later and is optimized for

Table 7.1
MPEG-2 Profiles and Levels

| | Profiles | | | | | |
Level	Simple	Main	4:2:2	SNR	Spatial	High
Low		4:2:0		4:2:0		
		352 × 288		352 × 288		
		4 Mbps		4 Mbps		
		I, P, B		I, P, B		
Main	4:2:0	4:2:0	4:2:2	4:2:0		4:2:0, 4:2:2
	720 × 576	720 × 576	720 × 576	720 × 576		720 × 576
	15 Mbps	15 Mbps	20 Mbps	15 Mbps		20 Mbps
	I, P	I, P, B	I, P, B	I, P, B		I, P, B
High-1440		4:2:0			4:2:0	4:2:0, 4:2:2
		1440 × 1152			1440 × 1152	1440 × 1152
		60 Mbps			60 Mbps	80 Mbps
		I, P, B			I, P, B	I, P, B
High		4:2:0				4:2:0, 4:2:2
		1920 × 1152				1920 × 1152
		80 Mbps				100 Mbps
		I, P, B				I, P, B

Reproduced with permission from [1].

use at higher data rates and higher resolutions to provide for professional and high-quality home applications. The options of MPEG-2 are defined by *profiles* and *levels*, which provide choices from low resolutions of 352 × 288 up to HDTV resolutions of 1920 × 1152. Data rates go from 4 to 100 Mbps. The terminology is to specify "profile at level" (e.g., Main Profile at Main Level [MP@ML]). The choices are shown in Table 7.1.

The MPEG group is working on additional MPEG standards for other applications.

7.5 AUDIO COMPRESSION

Compression of audio was developed much earlier than video compression; that was because of the lower bandwidth or data rates of audio and because of its significance to the telephony market. In that field, compression was even done with analog techniques before the general availability of digital hardware.

Audio compression is seldom used in professional recording. That is because uncompressed, high-quality audio data rates are small compared even to highly compressed video data rates, so the small improvement in total audio/video data rate is not worth the complexity, performance trade-offs, and cost of compressing the audio. However, it is used in computer audio and is part of the digital video broadcasting standards.

7.5.1 Basis for Audio Compression

Audio does not contain the same degree of redundancy as scanned video. That is beacuse an audio data stream has no equivalent to the video scanning structure, from which much of the video redundancy is obtained. Most audio compression is based on nonaudibility

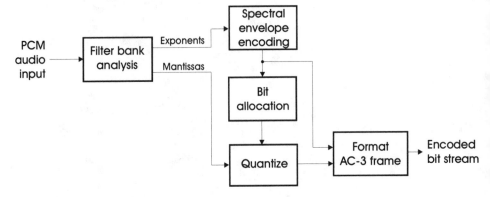

Figure 7.8 *Block diagram of AC-3 audio compression. Reproduced with permission from [1].*

(equivalent to nonvisibility in video), which takes into account the ear's poor response to high frequencies and the effect of multiple sounds affecting the audibility of each other.

7.5.2 Audio Compression Techniques

Any of the techniques described in Section 7.4.3 can be applied to an audio stream, but they are not very effective because of the lack of any structure or patterns in an audio stream. The only one that has been widely applied is differential coding, since the assumption that changes between adjacent audio samples are small is generally valid. Nevertheless, as stated above, this breaks down when large sample-to-sample changes do occur. The problem is partially solved by making the differential coding *adaptive*—that is, the dynamic range of the difference coding changes according to the size of the current sample-to-sample differences. This is called *adaptive differential pulse code modulation* (ADPCM) and it provides up to 4:1 compression at moderate audio quality levels. Greater compression at high quality is achieved using transform coding methods based on the ear's response to different frequencies and amplitudes. The transforms for audio are very different from those used for video compression.

7.5.3 Audio Compression Algorithms

Three of the most important audio algorithms are ADPCM, Dolby AC-3 (now called Dolby Digital), and MPEG audio. These are discussed in this section.

7.5.3.1 ADPCM

As explained earlier, differential pulse code modulation (PCM) achieves its compression by encoding the differences between samples rather than the samples themselves. The compression is created because differences are usually small and can be encoded with fewer bits than the samples themselves contain. In a 16-bit per sample audio system, for example, differences might be encoded in as few as 4 bits. This works fine, except that the differences are not always small enough to be handled in only 4 bits (this would require that the differences remain less than 1/4,096 of the full 16-bit scale). Greater range can be obtained by encoding the differences more coarsely, but that reduces the amplitude resolution for small differences.

The solution is to make the scaling of the difference bits be *adaptive*, which means the scale is adjusted dynamically to best represent the current magnitude of changes. This is easily done during compression by looking at a group of samples and setting the difference scale to fit them. However, it is trickier to convey this information to the decoder so that the decoder knows what scale to use in reconstructing the samples. Of course, this could be done with an escape code to identify when scale measurements are being sent. Another way is for the decoder to calculate the scale from a group of reconstucted samples the same way the encoder does it. That approach requires no extra code assignments. However, there is still the problem that errors can propagate downstream if only differences are being used, since each sample depends on the previous one, and so on. This is the same as with video predicted frames. Periodically, an absolute sample must be sent to assure that errors will die out.

7.5.3.2 AC-3 Audio Compression

The audio system of the ATSC DTV [2] system has adopted the Dolby Digital (AC-3) system to provide high fidelity surround sound—five full-bandwidth channels (left, center, right, left surround, right surround) and a subwoofer channel. This is referred to as 5.1 channel sound, signifying the narrower bandwidth of the subwoofer channel. Using the compression, this audio is transmitted at a data rate of only 384 kbps, less than the data rate of a single audio CD (1.2 Mbps).

The encoding process for AC-3 is shown in Figure 7.8. Incoming digital audio for the six channels is blocked into 512-sample groups and transformed to the frequency domain by an algorithm referred to as the *filter bank*. The output of this process is in a special floating-point format using an exponent and a mantissa value for each coefficient. The exponents represent the spectral envelope of the audio and they are encoded that way. The mantissa values are subject to quantizing to reduced bits according to a method that considers the spectral envelope in choosing quantizing bits. The results of the spectral envelope and mantissa coding are assembled into a frame structure that holds the block contents for all six channels plus synchronization and identification words, so that the format is easily extracted at the decoder.

7.5.3.3 MPEG Audio

Both MPEG-1 and MPEG-2 standards include an audio component. MPEG-1 transmits either monophonic (one channel) or stereo (two channels) audio. MPEG-1 audio compression differs from Dolby Digital in that it divides the audio signal into 32 sub-bands and specifies appropriate sampling rates and quantizing levels for each sub-band. Based on a psycho-acoustic model, the sampling rates and quantizing levels may change dynamically as the audio signals vary. Compression ratios of 7:1 to 10:1 are achieved.

MPEG-2 extends this by matrixing as many as five channels into the two stereo channels, which are then compressed by MPEG-1 techniques. An MPEG-1 decoder can play this as two stereo channels, while an MPEG-2 decoder can unmatrix the information into the original five channels. MPEG audio compression is widely used in Europe and for satellite broadcasting. It is also one of the audio modes standardized for DVD-Video (see Section 11.2.5).

The MPEG-2 audio standards are widely used in satellite broadcasting and in Europe. In the United States, however, the ATSC DTV standards used MPEG-2 video and transport but with Dolby Digital audio compression.

7.6 TIME BASE CORRECTION

Video systems generally are designed with the expectation that scanning, subcarriers, and sampling will have precise and stable frequencies. This is easily accomplished in analog systems with crystal oscillators, frequency dividers, and so forth—that is, until recorders enter the picture. The output frequencies from analog recorder playback depend on the mechanical motion of the scanner and tape; although this can be servo controlled to have a precise average frequency, servos alone cannot guarantee that the frequency will be stable over short periods of time. The frequency has time jitter because of uncorrectable servo errors; this is known as *time base error.*

Video recorders were at the mercy of time base error until the *time base corrector*[1] (TBC) was developed. Without a TBC, recorders could not be fully synchronized with other devices in the studio, and their signals could not participate in video mixing or special effects, which are so important to modern video production.

The first TBCs in analog systems were expensive, difficult to manufacture, and had very limited range of correction. Now, with digital technology, a TBC is inexpensive and easy to implement for almost any degree of correction. Digital TBCs are described in this section.

It should be noted that home VCRs generally do not have a TBC but still produce stable pictures because of two factors: (1) the synchronization circuits of the TV receiver used for display follow the varying frequency of the recorder signal, and (2) the color-under system inherently provides a stable color subcarrier frequency. This is workable only because there is no requirement in the home to synchronize a recorder signal with other video sources. In a studio operation, all sources must be precisely synchronized so signals can be switched, faded, dissolved, or subjected to other effects. As the home recorder becomes digital, most will have TBCs because they are so easily included in a digital system.

---

1   The terms time base *corrector* and time base *compensator* are used in the industry. They both refer to the same device.

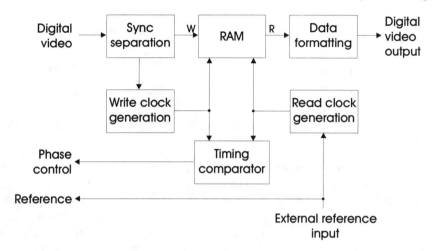

Figure 7.9 *Block diagram of a digital TBC.*

7.6.1 TBC Objectives

The requirement for a TBC is simple: insert a varying amount of delay into the recorder replay signal so every part of the output is precisely timed to a reference signal, which may be generated from a stable internal source or, most often, from an external sync generator.

7.6.2 TBC in Analog Systems

Today, there are no analog TBCs—all are digital. Analog systems use ADC and DAC to interface a digital TBC. This is less expensive and performs better than trying to do it all in analog.

7.6.3 TBC in Digital Systems

A block diagram for a typical digital TBC is shown in Figure 7.9. The synchronizing information is separated and a sampling clock is generated from the jittering bit stream to write the samples into a RAM memory. This memory may be large enough to store any-where from a few lines up to one or more whole frames of video. RAM has become so inexpensive that most systems have at least a frame of memory; this allows the TBC to be used for other features, such as still frame, slow-, or fast-motion playback.

The write clock and memory addressing must be positioned so the video scan lines are written at exact locations in the memory; thus, the memory will contain a copy of the video lines at fixed memory addresses—these can then be read out in synchronism with any external system by starting the reading of lines at the predetermined addresses. Thus,

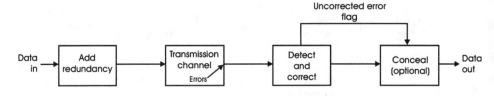

Figure 7.10 *General block diagram of a system with error protection. Reproduced with permission from [1].*

the read clock is derived from the external reference signal and reads the memory so that the output video will be synchronous with the external system.

The external reference also drives the servo system of the recorder to provide timing of the replay signal that accounts for the average delay through the TBC. By analyzing the relative timings of the write and read clocks, a feedback signal can be generated to adjust the servo timing to assure that the TBC memory will not overflow or underflow during operation.

This relatively simple architecture is readily implemented at low cost and is a part of all digital video recorders. Having the frame memory in the system allows many other recorder features to be provided.

7.7 DIGITAL ERROR PROTECTION

In addition to the capability of compression in digital systems, another major feature of digital systems is the ability to have nearly perfect error protection. Errors occurring for any reason in the record-replay process can be detected and corrected up to an error rate limit established when the system is designed. The processing for this can get very complicated, but with ASICs, it does not have to be expensive.

7.7.1 Basis for Error Protection

All error protection is based on adding redundancy bits to a data stream. By analyzing the data stream and the associated redundancy bits, the presence of errors can be detected and, in some cases corrected. Powerful algorithms have been developed for this process, as described below. Figure 7.10 shows a general block diagram of a transmission channel with error protection. Before passing through the channel (recorder), a controlled amount of redundancy is added to the data stream according to one or more of the techniques discussed below. This redundancy data is called *overhead* and it allows the later circuits to analyze the received data for detection and correction of errors. In many cases, more errors can be detected than can be corrected, but this offers the opportunity for further processing based on knowledge of the information content (scanned video samples, for example) to *conceal* any detected but uncorrected errors. Concealment means that something is done to make errors less visible (or less audible, for audio).

7.7.2 Techniques of Error Protection

Error protection systems generally do not need to know anything about the meaning or the information content of the bits they protect. However, they do need to know a great deal about the possible errors and their statistics. The *bit error ratio* (BER) is a key parameter; it specifies on average how many errors may occur in a given block of data, expressed as the ratio between the number of errors and the number of data bits in a given block. BER is expressed in negative powers of 10; for example, one error in a million bits is a BER of 10^{-6}. Error protection system designers also need to know how those errors occur—are they spread out all over the data or do they occur in groups of adjacent errors (*burst errors*)? From that information, the error protection architecture can be chosen, the block size and the number of redundant bits can be specified, and the method of analysis can be defined.

7.7.2.1 Parity

The simplest method of error protection adds 1 extra bit during recording, called a *parity bit*, to each small group of data bits (a byte, for example). That bit is defined by whether the number of 1 bits in the data group plus the parity bit is odd or even. In an *even parity* system, the parity bit is 1 if the sum of the data bit values is odd, and it is zero if the sum is even. An *odd parity* system reverses the definition of the parity bit.

After processing or recording, the system simply counts the number of 1 bits in the data group plus the parity bit to detect errors. In even parity, the sum of all bit values should always be an even number. If it is not, an error has occurred. This system is limited, however, to the detection of single errors in a data block and the errors cannot be corrected. All that is known is that a bad block has been received. An even number of errors in a data block is not detected at all.

Parity can be improved by viewing the data blocks as two-dimensional and adding parity bits in each direction (this is known as a *block code*). For example, 64 bits of data may be thought of as an 8 × 8 bit array as shown in Figure 7.11. An even parity bit is added for each row and each column, a total of 16 overhead bits in all. As the figure shows, a parity error in a row and another in a column will identify a single bit error. That can be corrected simply by reversing the value of the bit. However, this approach has 25% overhead (16/64) and it can correct unambiguously only one error per block. More sophisticated approaches can do better.

7.7.2.2 Reed-Solomon Coding

More advanced error coding becomes highly mathematical—far beyond what can reasonably be presented in detail here. However, some of the concepts will be discussed.

A block of data does not have to be viewed as a block of bits; the bits can be grouped into *symbols*, which are then capable of representing 2^N values, where N is the number of bits in each symbol. For example, symbols might be bytes. For a given block of symbols,

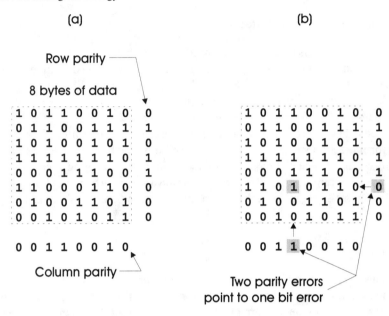

Figure 7.11 *Two-dimensional parity for correction of single-bit errors: (a) before and (b) after transmission. Reproduced with permission from [1].*

redundant symbols are added to the data stream for error correction (EC). These EC symbols are calculated from the data symbols according to an algorithm that will facilitate detection of faulty symbols on replay. It is this part of the system that is so difficult to explain, but we can view it simply as a mathematical process that generates EC symbols and adds them to the data stream. On replay, the data stream with the EC symbols is analyzed by a compatible mathematical process that produces no output when there are no errors, but when errors exist there are then two outputs: one or more *syndrome* values that indicate the locations of faulty symbols, and EC patterns that can be applied to faulty symbols for correction. In general, there will be more syndromes than patterns, meaning that some errors are detected but are not correctable.

The *Reed-Solomon* (R-S) codes are a group of mathematical processes that are extremely efficient in error detection and correction. They are specified in terms of the error correction block size in symbols and the number of information symbols in that block. The detection capability is approximately equal to the number of correction symbols, which is the difference between the EC block size and its information size. Thus, a 208,192 code could detect up to 16 faulty symbols in each 192-symbol block.

7.7.2.3 Interleaving

Recording systems are generally subject to burst errors because of local defects in the record medium. Bursts can be much larger than an error-correction block, often resulting in complete failure of correction. This can be overcome with the use of data *interleaving*

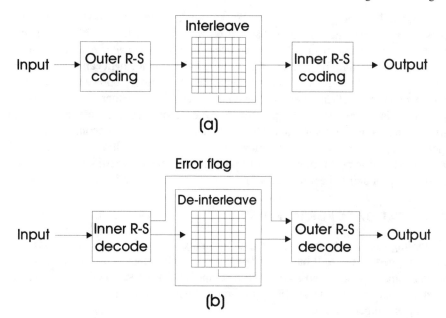

Figure 7.12 *Block diagram of cross-interleaved Reed-Solomon coding: (a) recording and (b) replay.*

between two R-S processes. This technique is called *cross-interleaved R-S coding* (CIRC) and is shown in Figure 7.12.

Interleaving takes a block of data and rearranges its order so that a burst error before interleaving, which would have a number of adjacent-symbol errors, is distributed to a number of single-symbol errors located throughout the data block. The usual method is to write a block of data into memory and then read it out using a different addressing scheme. This is done between two R-S encoders during recording and again between two R-S decoders on replay. The first R-S encoder is called the *outer* coder and the second R-S encoder is the *inner* coder.

When a replay burst error occurs, the inner decoder may be completely overloaded and cannot correct the block at all. An error flag is generated and passed by the de-interleaver to tell the outer coder that a whole block error has occurred. Because the interleave block size is much larger than the R-S block sizes, a faulty inner-code block will be distributed into a few faulty symbols in each of the outer-code blocks, where they are easily corrected. This architecture is capable of correcting burst errors of thousands of symbols [3].

7.7.2.4 Error Concealment

R-S coding operates without knowing anything about the format or information content of the data. It corrects errors up to its capability; beyond that, some errors can be located but not corrected. Using the error location information, further improvement is possible if the format and meaning of the data are known. For example, if each symbol in the data represents a sample value, a known bad sample may be made less visible by replacing it with a value calculated from adjacent samples. This is *error concealment*. It requires a knowledge of the data format so that the correct kind of replacement can be calculated for symbols known to be in error. This is practical when the symbols represent actual samples, but it becomes impractical when the data is compressed because there is no way of knowing what an individual symbol represents without decompressing the entire bit stream. Furthermore, a single bad symbol could cause a large defect in the decompressed picture, depending on where it falls in the compressed code. When handling compressed video, the error protection performance must provide an extremely low BER so that the uncorrected error probability becomes very low.

7.8 DIGITAL DATA FORMATTING

In recording, the bit stream to be recorded usually has to do much more than simply represent video or audio. It has to contain digital synchronizing information, identification data, auxiliary data, and so on. These components are assembled by formatting the data stream according to a specific algorithm that must be known in advance by both the recording and replay circuits.

7.8.1 Objectives of Data Formatting

The objectives of data formatting are as follows.

- *Combination of multiple input streams into one for recording in a single channel*—Most recorders have a single or, at most, a few recording channels, yet they must record a video signal, several audio signals, and possibly other kinds of data simultaneously. This requires means for interleaving all the signals into the available channels.
- *Synchronization*—The receiving device for a data stream must be able to detect the format structure of the data stream, either when the stream first begins or when it is entered at some point in the middle. This requires a means for synchronization.
- *Identification*—Since there may be different types of data in the same stream, means must exist for a receiving device to identify the currently flowing data.
- *Format options*—Data formats can vary within the same stream, so a formatting structure may need to provide information about its own format parameters.

The concepts of data formatting are very flexible and can be adapted to other objectives in different systems.

7.8.2 Data Formatting Techniques

The most common techniques of data formatting are discussed in this section.

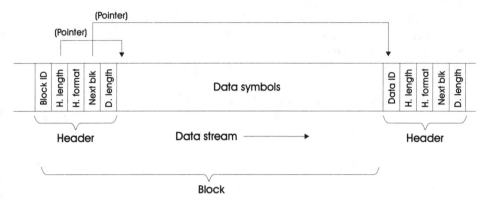

Figure 7.13 *A data block with a header.*

7.8.2.1 Synchronization Blocks

This is accomplished by inserting special symbols or blocks periodically to identify a particular place in the data. Such a synchronization (sync) block must have a unique pattern that can never appear in the data (or it must have at least a low probability of appearing in the data). Since synchronization must be completed before any other processing of the bit stream is possible, synchronization patterns cannot be compressed, interleaved, or changed in any other way before recording.

A data stream can be easily searched for sync blocks in real time by passing it through a shift register that has the same number of stages as the number of bits in the sync block. The stages of the shift register also each have a comparator circuit that compares the current bit in the stage to the corresponding bit of the sync word. On each clock cycle of the shift register, the comparators are also clocked—when all comparisons are true, the current data in the shift register is the sync block.

The data stream immediately following the sync block begins the format structure; usually the first thing is a header.

7.8.2.2 Headers

A data block that provides information about the data is a *header*. It has a predetermined structure that identifies the current data and its format. A header is not necessarily a fixed structure, it may be of variable length or have variable content. This is accomplished by having fields early in the header that indicate the header length and any header format variations. Of course, this part of the header must be fixed in format, but it can indicate variations in the remaining header structure. Figure 7.13 shows an example of a header containing some of these features.

Header fields generally indicate where the identified data begin by specifying the number of bytes or symbols from the start of the header to the start of the data ("H. length" in the figure). Such an indicator is called a *pointer* field. Another pointer often indicates where the end of the data is located, which usually would also be the position of the next header ("Next blk" in the figure). If the data identified by the current header are not to be used (skipped), the system knows how many symbols to skip before examining the next header.

Headers are usually preceded by sync blocks, although this is not always necessary if the data stream is not going to be entered randomly or edited. However, regular sync blocks allow the system to constantly check synchronization and respond by re-syncing if an error occurs.

7.8.2.3 Packets

A block of data that has its own header, and possibly a sync block, is called a *packet*. Packets can be of fixed size or variable length. Combining multiple data streams into one composite stream is easily accomplished by interleaving packets created from each of the input streams. Streams of widely differing data rates, such as audio and video, are easily combined this way. This was described in Section 1.6.2.6 regarding the MPEG data stream format.

7.9 CONCLUSION

Modern recorder signal processing is generally done with digital circuits, even when the record format itself is analog. That is because of the efficiency, high performance, and low cost of digital signal processing. As all record formats themselves become digital, the extra steps and degradation of ADC and DAC will be eliminated.

REFERENCES

[1] Luther, A. C., *Principles of Digital Audio and Video*, Artech House, Norwood, MA, 1997.

[2] ATSC Doc. A/52 (1995), *Digital Audio Compression Standard (AC-3)*, http://www.atsc.org

[3] Watkinson, John, *The Art of Digital Video,* 2nd ed., Focal Press, Oxford, 1994.

8

Editing

The process of assembling a program from separately recorded segments is known as *editing*. Editing, however, is only a part of the larger process of *postproduction*, which includes all the steps of selecting the exact shots to be used in a program, preparing them for use, and editing them together with special effects, transitions, or other means of assembly. Postproduction is related to *production*—the task of capturing all the shots that will eventually be used in a program (or might be used). Production provides the inputs to postproduction.

8.1 PRODUCTION-POSTPRODUCTION

The production-postproduction style of program creation first appeared in the motion picture film industry; with film, this method is mandatory because film is not a real time process—one has to wait for the film to be developed before assembling the program by editing the films together. Techniques are highly developed for film editing, for which a whole industry exists.

Electronic production-postproduction program creation was not possible in the early days of television because there was no way of recording TV pictures at that time. The only way to use film-style program creation was to shoot and edit the program on film, which was expensive and required converting the film to video at the time of airing. Most early TV programs were created and assembled in real time as they were being shot, which seriously limited the kinds of shots, the quality, and the special effects of early TV programs. The TV industry eagerly awaited the development of video recording so that film-style program creation, with its greater flexibility and capability, would become possible.

When videotape recording was first introduced by Ampex in 1956, most people viewed it as a tool for delaying broadcasts so they could be played 3 hours later for the West Coast. Thus, the market was seen as very small—not many people were doing delayed broadcasts. The TV programming community, however, saw it differently and immediately embraced the technology as a production tool. In the early days of videotape, editing was done only by cutting the tape and rejoining it—a mechanically tedious process.

Technology for electronic editing was soon developed and the videotape postproduction community was off and running. Today, essentially all professionally created video programming is created by the production-postproduction method. As this technology has matured, it has become both more capable and lower in cost—thus supporting not only very sophisticated professional postproduction, but also less sophisticated systems for semi-professional and home users.

8.2 EDITING PRINCIPLES

Editing for program assembly requires three steps.

1. *Shot selection*—This is the process of reviewing the production material and selecting the exact pieces (shots) to be used in the program. The shots will be assembled in a specified order using one or more *transition effects*. Each junction between shots is called an *edit*.

2. *Edit previewing*—It is desirable to view the edits in a preliminary manner before completing the final assembly. Different editing systems have more or less capability to do this. During preview, the shot selections may be modified to optimize the overall effect of the assembly.

3. *Final assembly*—This is the process of actually implementing the edits to produce an *edited master* version of the program.

The facilities required for these steps are discussed below.

8.2.1 Time Coding

Editing requires precise identification of video frames, so that shot locations can be unambiguously identified. The normal approach is to sequentially number frames as they are recorded; generally, the time of day is used in the numbering scheme. Time of day is not necessary for editing, but it is a convenience during production because events in the recording can be noted simply by consulting a clock. In all but the most rudimentary systems, frame numbering is accomplished by including a *time code* with the recording. Alternatively, a time code can be added after the original recording.

In magnetic recording, the most widely used time code is the *SMPTE time code*, defined by SMPTE Standard 12M [1]. This may be recorded on a separate longitudinal track, which is convenient for reading while tape is being shuttled at high or variable speeds. This is called a *linear time code* (LTC). An alternative is to include the time code data in the video signal, such as *vertical interval time code* (VITC). Other time code formats are used in some home recorders.

In digital recorders, a time code may be included in the frame header structure of the video data. This is readable any time the video data are being read. Some digital recorders also have a longitudinal time code track; this is provided because it can be read even when video is not being read.

For shot selection, a special copy recording is sometimes made with the time code information added right in the video picture. This is called *burned-in time code*, which allows shot selection to be done on almost any player that has the ability to do still-frame and variable-speed playback. Home recorders are often used for this purpose—the production people can take burned-in tapes home and perform preliminary shot selection.

When audio is recorded separately from the video, it also must be time coded so that proper audio-video synchronization can be achieved in editing.

8.2.2 Edit Decision List

A list of the details for all the edits in a program must be created and maintained during editing. This is used for previewing edits and again in the final assembly of the program. Such a list is called an *edit decision list* (EDL) and it includes time code locations for all the video or audio components involved in each edit, as well as transition instructions and any other parameters.

EDL formats may be specific to the editing system being used; however, SMPTE has written Standard 258M for exchanging EDLs between systems of different manufacturers [1]. Figure 8.1 shows an example of an EDL for some simple edits; this roughly follows SMPTE 258M. The timeline at the top of the figure shows the edit sequence. The table

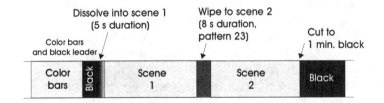

Edit Decision List

Edit No.	I.D.	Mode	Effect	Source Entry	Source Exit	Sync Entry	Sync Exit
01	CBARS	B	C	00:00:00(0)	00:00:00	00:00:00	02:00:00
02	BLACK	V	C	00:00:00(0)	00:00:00	02:00:00	02:15:00
03	BLACK	V	D 150	00:00:00(0)	00:00:00	02:15:00	03:00:00
03	SCENE1	V	D 150	05:13:22	05:58:22	02:15:00	03:00:00
04	SCENE1	V	W23 240	05:58:23	08:28:23	03:00:01	05:30:00
04	SCENE2	V	W23 240	01:30:00	04:00:00	03:00:01	05:30:00
05	BLACK	V	C	00:00:00	01:00:00	05:30:01	06:30:00

Figure 8.1 Example of a simple edit decision list. Reproduced with permission from [2].

below is the actual EDL, one or more lines per edit. Individual edits are identified by their edit number; multisource edits have more than one line to define the sources. The mode field is coded A for audio-only, V for video-only, or B for both. The effect column defines the type of effect and its parameters: C is for cut, D is for dissolve, W is for wipe (there are many other transition types—see the next section). The source time code columns specify the start and end time codes in the source reel, and the sync fields specify the start and end time codes in the target recording. The color bars and black signals are still pictures, which is indicated by the (0) after their entry time codes. The dissolve in edit 03 has a duration of 5 seconds (150 frames), and the wipe in edit 04 is wipe pattern 23 (a rectangle opening from the top of the picture) with a duration of 8 seconds (240 frames).

8.2.3 Transition Effects

There are many ways to join two scenes during editing; these are called *transitions*, or *transition effects*. The simplest transition is to switch between signals during the vertical blanking interval—this is a *cut* transition and it can seem abrupt to the viewer when the signals are very different. Many effects are available to smooth the transition and make it less abrupt; some of these are shown in Figure 8.2.

Probably the smoothest transition is the *dissolve*, where the amplitude of the current signal is reduced while simultaneously increasing the amplitude of the new signal. At the end of the dissolve, the new signal has replaced the old signal. A relatively fast dissolve (less than 1 second) smooths a cut, but a slower dissolve causes the transition itself to be noticed by the viewer. Many of the fancier transitions call attention to themselves, which may actually interfere with the visual flow of the program.

Many dynamic transitions are in the category of *wipes*, where a moving transition line cuts the new scene into the previous one. These come in various patterns, a few of which are shown at the bottom of Figure 8.2.

Except for cuts, transition effects require the availability of both the previous and next scene simultaneously during the effect. An electronic switching box performs the transition and outputs a single signal that contains the effect. Thus, recorders alone cannot do dynamic transitions—it takes a video switcher.

Some special effects are achieved by processing the signal itself, as shown at the top of Figure 8.2. These effects are generally done digitally and may be used as transitions or as single-signal effects. The generation of such special effects was one of the first applications of digital technology in video systems.

8.2.4 Edit Previewing

The process of shot selection requires the ability to play each of the shot recordings and select the exact frame locations to begin and end each transition. Once this has been done, it is generally necessary to preview the actual effect to check that the selected frame locations produce the desired visual effect. This requires running the recordings and the video switcher or effects unit under control from the EDL. Since the preview process may

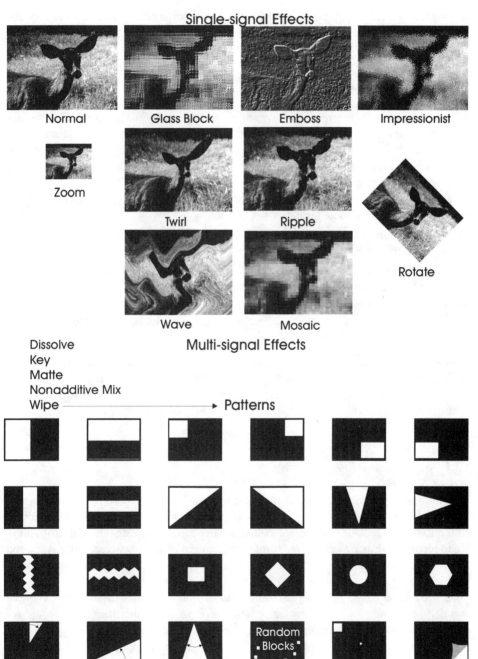

Figure 8.2 *Video transition effects. Reproduced with permission from [2].*

be done many times in optimizing a single edit, it is important that previewing be easy and fast. This works better with disk recorders rather than tape recorders because of the faster random-access capability of disks.

8.3 LINEAR EDITING—VIDEOTAPE

To get from one point in a recording to another, the tape between those points has to pass through the tape deck. This is because a tape recording is *linear*—one-dimensional along the tape. Disk recorders are different—they can jump almost instantaneously from one part of the recording to another simply by moving their heads along the radius of the disk—thus, they are called *nonlinear*.

Since editing involves joining scenes that are recorded at different places on the tape, tape's linearity causes some limitations in editing.

- Multiple source tape decks are needed so that scenes can be positioned to play sequentially regardless of where they are on the source tapes.
- Shot selection may be awkward because of the mechanical behavior of the tape deck when shuttling or still-framing tape.
- Edit previewing may be slow because of the need to run tape to set up the timing for each transition. Without special preparation, it may not be possible to preview more than one edit at a time (see Section 8.3.2).

In spite of the limitations, until recently, nearly all programs were assembled by videotape editing. Disk-based editors have existed for years but they were expensive and had their own set of limitations. In the past 5 years or so, the limitations and high cost of disk-based editing have been overcome, and that is now the preferred approach for program creation (see Section 8.4).

8.3.1 Assemble and Insert Editing

The simplest form of tape editing uses two tape decks—one for recording the edited master and one for playing source tapes. This is based on the *electronic editing* capability built into most videotape recorders to switch instantaneously (during one VBI) from playing to recording or back again. This feature can be used to add a new recording at the end of an existing recording (*assemble editing*), or to overwrite a section of an existing recording (*insert editing*).

Since there is only a single signal source and no video switching capability in a simple two-deck configuration, this approach cannot perform transition effects that would require combining two signals during the transition. Electronic editing is basically cut-only. Some home decks are able to play and record simultaneously, so the playback signal can be mixed with a new signal and recorded back on the same tape, creating effects with a built-in effects processor. With analog recorders, it is difficult to obtain high-quality performance from this mode. With digital recorders, simultaneous read-write is practical because the digital error correction takes care of any crosstalk signals. This mode is still not very popular, however, because it restricts the transition effects to those built into the recorders. Most users enjoy the flexibility of a separate effects unit in their editing system.

8.3.1.1 Basic Capabilities for Electronic Editing

The basic capabilities required for a videotape recorder to do electronic editing are as follows.

- VBI switching of record current to the helical heads, synchronized with the tracks already on tape;
- Flying erase heads on the helical drum, activated with proper timing to erase helical tracks before recording;
- Switching of the control track from playback to recording timed with the edit;
- Synchronization so that new recorded tracks are properly positioned relative to the previous recording;
- Similar capabilities for audio tracks, controllable with the video or separately.

Most recorders have controls for electronic editing accessible from the front panel. Recorders designed for use in editing systems also have remote control of the editing features so that an external edit controller can operate the recorder.

8.3.1.2 Assemble Mode

Figure 8.3(a) shows the operation of assemble mode. In this mode, a new recording extends beyond the end of the previous recording. The recorder is put into playback mode before the edit point so that the scanner is synchronized with the existing helical tracks. At the point of switching to recording, the capstan servo must transition from a tracking servo to a speed servo. This transition must be smooth so that subsequent playback through the transition region will not have any disturbance. Longitudinal tracks also are switched to recording at the edit point.

Programs can be created by repeated assemble recordings; however, this is awkward when it becomes necessary to make changes in the recording—everything after the change must be rerecorded.

8.3.1.3 Insert Mode

A new recording can be made to overwrite an existing recording by using the insert mode. This requires an existing recording to be present all the way to the end of the new recording. That being the case, the capstan servo can be left in control track playback mode while recording; this will keep the tracks synchronized with the existing tracks so there will be no disturbance at the end of the insert recording.

By beginning with a tape prerecorded with black picture and time code, programs can be assembled in insert mode. This has the advantage over assemble mode in that scenes

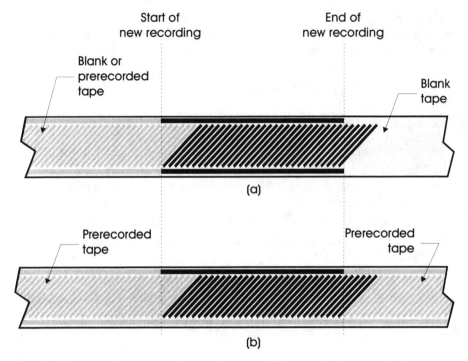

Figure 8.3 *Videotape editing modes: (a) assemble edit and (b) insert edit. Reproduced with permission from [2].*

can be recorded in any order, placing each one at its proper location according to the time code track. If a scene must be redone, it can be overwritten again using exactly the same time code values. Of course, the total length of recording cannot be changed—if one segment has to become longer, another has to become shorter.

8.3.1.4 Synchronization for Electronic Editing

Analog helical recorders are generally synchronized to switch tracks during the VBI and have one field per track. Track switching occurs at the specific angular drum positions where each head begins reading a track (near the edge of the tape). In editing service, the servo for the scanner drum is phase locked to the vertical sync of the incoming signal so that track switching occurs during the VBI of that signal. To assure proper tracking, the recorder is kept in playback mode before the edit, which means that the capstan servo is controlling tape speed according to the control track playback.

Digital recorders may not have the field-per-track configuration; they may be segmented, having several tracks per field or, with video compression, there may be several fields per track. This complicates the synchronization problem and generally requires decoding of the digital format to locate edit points. With compression that uses motion compensation (see Section 7.4.3.2), editing is further complicated because individual frames may not be independent of one another. In such a system, editing is generally only possible on intracoded-frame boundaries.

For frame-accurate editing, the recorders involved also must have time code synchronization. Since each recorder in general is operating with different time code values, the frame synchronization system must provide for the precise difference in time values to achieve the proper frame relationship at the edit point. That is usually accomplished by first achieving servo synchronization and then checking time code values; if the match is not correct, synchronization of one or both recorders is slipped in the proper direction to establish the proper time code relationship.

To allow time for servo and time code synchronization before an edit, the tape must be started a few seconds ahead of the actual edit point (called *pre-roll*); the amount depends on the synchronization-time specification of the recorder. Edit controllers automatically account for this, so the process is invisible to the user except that it makes editing (or previewing) take longer.

8.3.1.5 Timing in Electronic Editing

Timing issues in editing are handled by both the recorder and the edit controller. In general, the edit controller watches time code values, issues commands to the recorders for pre-roll, and starts recording when the appropriate time code value appears. The recorder manages its internal timings in response to a single "record" command to begin recording the frame indicated by the time code. The operation of an edit controller is discussed in Section 8.3.3. Edit timing issues in the recorder include the following:

- Control of the flying erase head—following a "record" command, the erase head must be activated when it begins passing over the first track of the section to be recorded. That corresponds to a specific angular position of the scanner drum, which is calculated by using a pulse from the scanner tachometer wheel and counting or delaying to reach the exact timing. It is usually necessary to switch off the erase head when it leaves the video portion of the helical track to prevent it from affecting any of the longitudinal tracks that the scanner heads may pass over. Thus, the erase head current switches on and off for each track of the edit.

- The recording heads are positioned behind the erase heads on the scanner drum; thus, they reach the start of the helical track later than the erase heads, and there are different time settings for switching record head currents. Again, this is calculated from the scanner tachometer pulses, and recording is activated on the first complete helical track after receiving the "record" command.

- When longitudinal track audio is being edited along with the video, similar timing control of the start of audio erasing and recording is required. A delay will be required between the application of audio erase current and the start of actual audio recording; this is determined by the distance between the audio erase and recording heads and the tape speed.

- In assemble mode editing, the control track also must be recorded for an edit. Control tracks are usually saturation recordings and are therefore self-erasing. Thus, the control track switches to recording at the start of erasing the first video track. Of course, the capstan servo must switch from playback operation to its coasting mode before recording the new control track. In insert mode, the control track is not rerecorded.

- A special situation occurs in handling a time code track for assemble editing. Usually time codes are determined by a time-of-day clock and are recorded continuously so that time code values progress smoothly along the tape. Edit controllers depend on this. In assemble editing, the time code generator's time value must be synchronized with the tape playback ahead of the edit so that it will provide a smooth sequence of values when the system switches into recording.

Since all these timing issues are specific to a particular tape deck design, they are always considered to be internal functions of the recorder and do not concern either the edit controller or the user.

8.3.1.6 Electronic Editing of Audio

Audio editing is performed either at the same time as video or separately (audio-only editing). Because there are no frames or fields in analog audio, the starting and stopping of edit recording must be done carefully to avoid audible signal disturbances at edit points. Various strategies have been used for this in analog audio recording; they generally involve a short dissolve between the existing recording and the new recording, with erasing coming on during the dissolve.

Digital audio is generally divided into blocks, often corresponding in duration to the video frames. Thus, editing of digital audio can be handled with cut transitions between blocks in the same way as for video. Most digital formats (see Section 10.2) record audio data at specific locations along the helical tracks; this is necessary because the bandwidth of longitudinal tracks is generally not high enough for digital audio. Helical-track audio is also mechanically simpler and costs less since separate audio heads are not necessary.

Record switching of the audio blocks in helical tracks is handled by switching at specified angular positions of the scanner drum the same way as for video. The design of the record format must provide guard bands between audio and video blocks along the track to allow for tolerances in the record switching.

8.3.2 A-B Roll Editing

A most effective method of videotape editing involves the preparation of two (or more) intermediate videotapes that contain alternate scenes, as shown in Figure 8.4. This is known as *A-B roll editing*, where the two tapes are the A and B rolls. The actual edits are made by running the A and B rolls in synchronism, and an edit controller operates a video switcher and effects unit to assemble the scenes in real time. A third VCR records the output of the switcher, which is the final program.

Although A-B roll editing is very flexible, it is extra work to prepare the intermediate tapes, which represent another generation of recording. Thus, the edited master is a third-

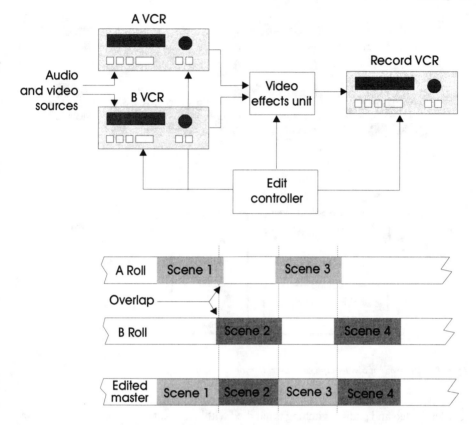

Figure 8.4 *A-B roll videotape editing. Reproduced with permission from [2].*

generation recording. If copies are made from that for distribution, the result is fourth-generation recording. In analog systems, this causes accumulated distortions. Of course, that is not a problem with digital recorders.

8.3.3 Edit Controllers

Tape edits are done one at a time. Edit controllers can be designed to handle only one edit or they can remember an entire EDL, so that previous edits can be revisited if necessary. This section describes an edit controller of the former type—one at a time editing.

Figure 8.5 is a diagram of such a simple tape edit controller that operates with two VCRs, one for a signal source and one for recording the edited master. Alternatively, the editing signal source can be a video camera, which allows live program assembly onto tape. Simple programs can be assembled this way in the field, but it offers limited flexibility compared to postproduction editing.

The edit controller is shown as its control panel; it is connected to two VCRs that have remote control, time code outputs, and video monitors that show the outputs from each

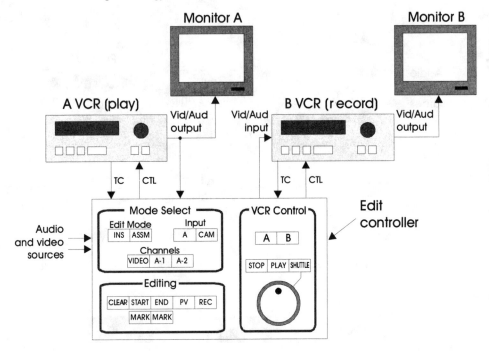

Figure 8.5 *Diagram of a videotape edit controller for one-at-a-time editing.*

recorder. Video and audio switching is handled within the edit controller; this is necessary for the edit preview function.

The editor control is divided into three sections. One provides for edit mode selection, including selection of insert or assemble editing, input selection between VCR or camera, and choice of video or audio editing modes. The second area of the editor panel controls the edit preparation sequence, taking the user through the sequence described below for setting up an edit, previewing it, or revising it. The third control panel area is a VCR control panel that can be switched to manual control of either VCR. This is used for finding exact frames for the edit actions.

The operation of this editor is as follows.

- A new edit is begun by pressing the Clear button, which clears the time code memory.

- The in-edit point on the master tape is determined by manually positioning the record machine (B) to the exact frame to begin the edit. This is found by using the VCR controls for recorder B in shuttle-jog mode until the exact frame is displayed on that recorder's monitor. When the desired frame is shown, the Start Mark button is pressed.

- The in-edit point for the source machine is determined by manually positioning the playback machine (A) to the exact frame as shown on its monitor, using the VCR control panel switched to recorder A. This is entered into memory by pressing the Start Mark button again.
- At this point, the in-edit can be previewed by pressing the Preview button, which will automatically pre-roll the recorders and run them through the edit point. Since no out-edit has yet been defined, the recorders must be manually stopped after viewing the in-edit.
- If the preview requires modification, the appropriate tape is repositioned and the Start Mark button is pressed again. This and Preview can be repeated as much as necessary to achieve the desired edit effect.
- If an out-edit is required, the End button is pressed and the same procedure is followed to determine it and preview it along with the in-edit. Note that the out-edit is determined only on one recorder, since the relationship between the two recorders is already determined by the in-edit point.
- When preview of the entire edit is satisfactory, it is recorded to the master tape on recorder B by pressing the Record button. The final edit can be viewed by manually controlling recorder B.

Note that there is no time code display on this editor. It is not necessary for one-at-a-time editing. In fact, such editing can be done without a time code at all; some low-cost VCRs have the capability to record editing marks on a longitudinal track, such as the control track. These are used to mark in and out editing points or simply to provide bench marks from which an edit controller can count frames to edit points. However, this approach cannot be relied on for frame-accurate editing because there is no way to obtain time code synchronization after the recorders are started for preview or edit recording. Frame errors of one or two frames can easily occur.

Much the same procedure is used to operate the editor for an audio-only recording. It is only necessary to choose the appropriate tracks to record using the buttons in the mode selection area of the panel.

The edit system shown in Figure 8.5 requires at least five separate equipment units. These can be combined in a single dedicated editor, like the Sony Betacam SX™ Digital Portable Editor described in Section 8.6.1.

8.4 NONLINEAR EDITING—DISK

As explained at the start of this chapter, operation of a videotape editor is slowed by the requirement of tape shuttling and pre-rolling for previewing or editing. These problems are overcome if all the material is stored on disk recorders, which can be rapidly random-accessed to display any point in the recordings. This is *nonlinear editing*, which is the preferred method for professional editing.

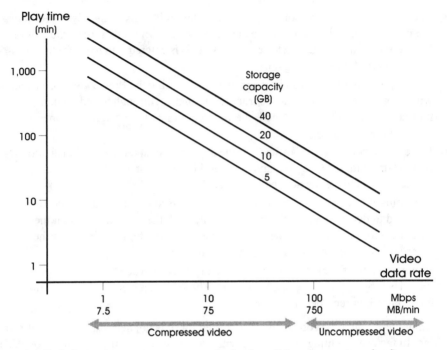

Figure 8.6 *Disk playing time versus storage capacity (neglects disk management overhead).*

8.4.1 Disk Capabilities

With the exception of the Laser Videodisc, all disk recorders for video are digital. This is necessary because of the limited recording area of a disk; only with digital video compression can a practical amount of video be recorded on a reasonable-sized disk. (Some high-end professional nonlinear systems do not use compression, or only very little—as the cost and capacity of disk recorders improves, there may be more systems like that.)

Small digital disk drive capacities have been steadily increasing—at this writing, single-drive capacity has passed 20 GB. That is just a point on a curve that will keep going up in the future. Figure 8.6 shows video recording time versus data rate and storage capacity. It shows that respectable playing times are available with 20 GB of storage for reasonable amounts of compression and that uncompressed storage is around the corner for multiple 20-GB drives. Thus, disk-based video storage and editing is becoming more practical all the time.

Another capability of disk recorders is important for editing—access time. In general, video is recorded on digital disks without regard to the relationship between video frames and disk tracks. Although the disk itself is capable of random access in a matter of milliseconds, video frames cannot be rapidly located without a table of their physical disk addresses. Lacking such a table, it would be necessary to read a video track until the desired frame number was found, which is a linear operation that does not exploit the nonlinear capabilities of the disk. Thus, a frame locating table should be created as the disk is recorded. This is easily done if all frames have the same data size; it is more complex if the compression used does not give equal-size frames.

Although disk access time is fast, it is not fast enough to start retrieving a new video during the VBI, as would be required for a cut transition. This limitation is easily overcome, however, by buffering small amounts of video in system memory. Data can be read from the memory buffer while the hard disks are seeking the next video segment.

8.4.2 Control of Disk Editing

The operation of a disk editor is much the same as was described for tape editing in Section 8.3.3. The difference is that there is no waiting for tape rewinding or pre-roll and multiple edits can be previewed almost instantaneously. The edited master for a nonlinear editor is simply the EDL—it can be played at any time for review or for recording to tape. A nonlinear editing system is described in Section 8.6.2.

8.5 PICTURE QUALITY ISSUES IN EDITING

Analog editing requires recording and rerecording of the signals, often three or more times. That causes multiple generation accumulation of errors. Digital systems can avoid this problem; nonlinear editing does not build up generations at all because the edited sequence is played directly from the original disk recordings, no matter how complex the editing. There can still be video distortions in digital editing, however, when lossy compression is used. That is because compressed signals generally have to be decompressed to video components to perform video effects, even something as simple as a wipe. If a lossy decompression-compression process is repeated, the distortions caused by the compression compromises can accumulate.

8.5.1 Multiple Generations

As has been shown, multiple generations of recording inherently arises in editing of analog recordings; with simple editing, an edited master is second generation, and distribution copies are third generation. More complex effects are possible when the second generation masters are used as sources for additional edits or effects, which are then recorded, producing third generation masters. The fancier one gets with the effects, the more generations may accrue. The designers of analog recorders had to adopt specification objectives that allow for the degradation of multiple generations yet still produce acceptable picture and sound quality after editing.

This build-up of generations and associated picture and sound degradations are one of the strongest arguments for editing digitally, even when the source materials are analog. In the digital case, the only degradation occurs in the original ADC and final ADC—as long as everything remains digital within the editing process, degradations cannot accumulate. The same situation is also a strong argument against the use of lossy digital compression in an editing system. In spite of the proponents of compression, the professional editors know what they are talking about regarding the effects of compression in editing and are often prepared to pay the cost of uncompressed storage. Since that cost is steadily dropping, this argument may eventually become moot.

8.5.2 Frame Accuracy

When editing a particularly critical transition, the difference of one frame in edit positioning can have a large effect. Thus, precise frame accuracy and repeatability are essential to professional editors. Systems having time code and time code synchronization are capable of complete frame accuracy. Editors that do not use time code generally do not have precise frame accuracy—there may be a tolerance of plus or minus one or two frames in repeated previews or running of the same edit. This is only acceptable in noncritical situations and has been relegated mostly to the home market. Even there, the use of time code is growing.

8.5.3 Picture Stability at Transitions

At the point of an edit there may be a picture disturbance because of a slight error of synchronization between the two signals being joined on the edited master. This can cause display synchronization to jump either horizontally or vertically or to show mismatch between signals that are combined in a dynamic transition. A similar problem may occur because of mistiming the erasure; starting erase too soon will produce a slight signal interruption that may disturb synchronization of a playback VCR or a display. Both of these problems are properties of the electronic editing circuits of the recording VCR. They are generally not an issue in digital tape recorders or disk recorders simply because precise synchronization occurs almost automatically in a well designed digital system.

8.6 SPECIFIC EDITOR DESIGNS

This section gives some examples of current editing system designs. They represent a range of capabilities and are not necessarily the ultimate; in fact, as with everything else in video systems, editor design steadily advances—what is the ultimate today becomes ordinary tomorrow and the frontier keeps advancing.

8.6.1 Sony Digital Portable Tape Editor

Figure 8.5 showed a simple editing system that requires several separate units connected together. This is not so bad in a fixed and dedicated installation, but it is awkward for field use. It used to be that no one attempted editing in the field because it took so much equipment and skill, but that is changing, and with the possibility of in-field editing, many producers have seen considerable advantages. Seeing the edited program, or at least part of it, before ending a field shooting session allows retakes to be made for problems that would have been very difficult to correct back at the editing suite. Sony has designed a portable editing system for field use—their DNW-A25/A225/A220 series.

Figure 8.7 *The Sony DNW-A225 Digital Portable Editor. Photo courtesy of Sony Corporation.*

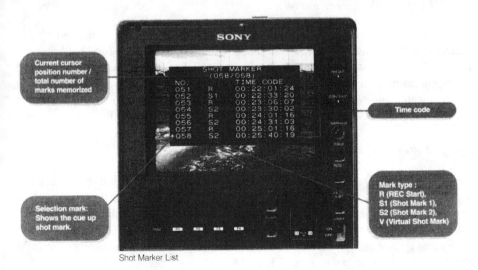

Shot Marker List

Figure 8.8 *Shot marker list display of the Sony DNW-A225 Digital Portable Editor. Photo courtesy of Sony Corporation.*

The basic idea is to have two VCRs, two monitors, and an edit controller assembled in a single portable package, as shown in Figure 8.7. The two recorders are actually separate units that can be used independently but they are coupled together in the field package to make the convenient field editing system. Operation is similar to that described in Section 8.3.3 except that a "marker list" is kept by the system. Instead of the single tape deck control panel shown in Figure 8.4, each recorder has its own shuttle-jog control. Other controls provide for capturing time codes (markers), previewing, and making adjustments to edit points.

A list called the Shot Marker List is kept by the system; this is not quite an EDL—it is simply a list of time code locations that can be selectively accessed during editing. The markers are helpful in setting up edits. Markers can be created during production, if desired, because the Betacam SX format provides for markers to be recorded on tape during shooting to identify key points in the material to be accessed during editing. When a tape is placed in the editor, it can be played through to create a marker list of time codes in memory that is then used to random-access to any of the markers. Additional markers can be added during editing or existing markers can be adjusted by previewing edits. Figure 8.8 shows how the shot marker list appears on the editing screen.

8.6.2 Nonlinear Editor

Nonlinear editors are available as turnkey complete systems, or they can be assembled using PC components and software. In either case, the operation is similar to the tape editing systems already described except that it is much faster and has additional capabilities.

8.6.2.1 PC-Based Nonlinear Editor

Figure 8.9 is a block diagram of a PC-based nonlinear editing system. The architecture is a conventional PC bus structure, which supports control and display for the editing system. A central mass storage unit holds programs, EDLs, and audio or video archives. Video processing is in hardware, including a compression unit, video effects and mixing, and a dedicated audio/video bus. This relieves the CPU from most of the video processing; such video processing units are available as add-in cards for the PC bus.

8.6.2.2 Professional News Editor

The Sony DNE-1000 is designed for editing news programs, although it is certainly suitable for other uses as well. It has its own dedicated PC and a separate chassis containing hard disk storage and video and audio processing circuits. It may also operate with an external video server for storage. Video is handled in MPEG-2 4:2:2 Profile @ML, which is the same format used in the Betacam SX (see Section 10.2.6). Operation is controlled from a graphical screen and keyboard. Figure 8.10 shows a screen of this editor. There are three main windows.

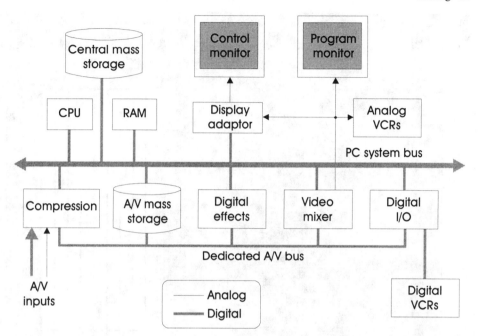

Figure 8.9 *Block diagram of a PC-based nonlinear editing system. Reproduced with permission from [2].*

1. *Viewer window*—where selected video is previewed for selection of edit points. A slider allows direct positioning of the place in the video to view, and VCR-type controls are below for controlling playback functions in shuttle or jog. In- and out-point selections show in two small windows to the right.

2. *Logging window*—shows all the shots that have been defined for this session, each in its own small window. Shots can be edited by moving them to the viewing window for adjustment.

3. *Program window*—shows the program in timeline form. As shots are selected, they can be moved into the program window to assemble the program sequence. Video transition effects are defined in this window by calling up a separate window for this purpose. Separate audio or video editing, or video level adjustments can also be done from this window.

A nonlinear editor is not restricted to one edit at a time but can play through an entire EDL without interruption. Thus, any part of the program being assembled can be instantly previewed. When editing is completed, the entire program can be played to a video recorder to create a master tape for distribution. By saving the source video clips and the EDL in digital form, the program can be reloaded into the editor at any time for review or modification.

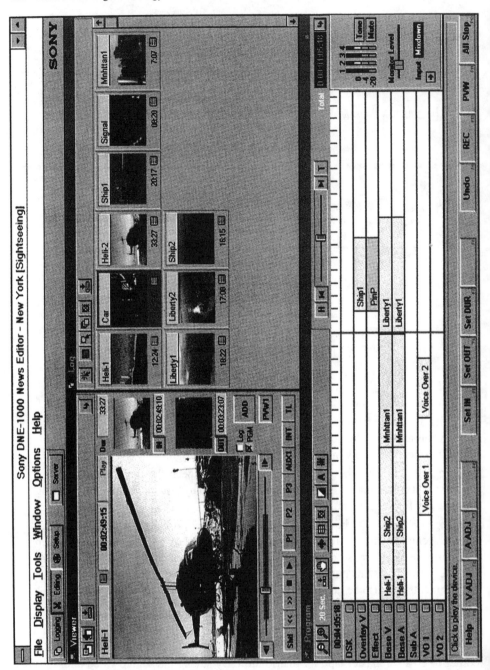

Figure 8.10 Nonlinear editing screen of the Sony DNE-1000. Photo courtesy of Sony Corporation.

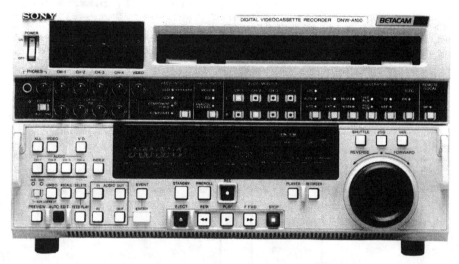

Figure 8.11 *Photograph of the Sony DNW-A100 Digital Video Hybrid Recorder. Photo courtesy of Sony Corporation.*

8.6.3 Sony Digital Video Hybrid Recorder

Most production shooting is done with camcorders and tape recorders. Users of nonlinear editing always face the problem that production tapes must first be played into the system and stored on hard drives before any editing can be done. Thus, tape players are always a part of nonlinear editing systems. Sony recognized this and designed a unique product in the Betacam SX line, the DNW-A100, that combines a tape recorder and a hard disk recorder into a single unit. The unit is capable of high speed transfer of tape to disk at 4× normal tape speed, and nonlinear editing can be done within the unit. For more advanced editing features, the unit may be connected to an external nonlinear edit controller. Figure 8.11 is a photograph of the product.

The use of this product is shown in Figure 8.12. Tapes from production are run in the tape deck of the DNW-A100 for transfer to the internal hard disks. With Betacam SX tapes, this can occur at up to 4× normal speed. (Betacam SP analog tapes can transfer only at normal speed.) The high-speed tape playback is achieved by a unique design of the helical scanner that has four sets of playback heads. With the scanner running at normal rotational speed and the tape running at 4× speed, the four heads are able to pick up all the data, which are assembled into memory by the signal processing circuits. The resulting bit stream is recorded on the hard disks.

A built-in editing controller allows nonlinear editing right in the unit, although an external edit controller can be connected for more sophisticated operations. Assembled programs can be output from the hard disks at 4× speed for transfer to a video server, or at normal play speed for transfer to videotape.

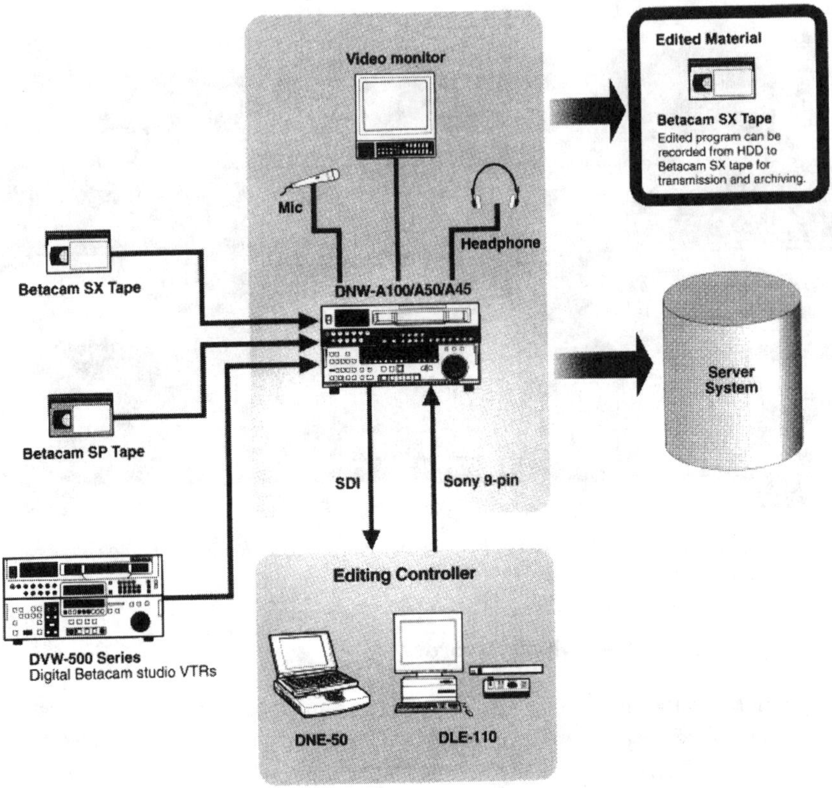

Figure 8.12 *Operating diagram for the Sony DNW-A100 Digital Video Hybrid recorder. Photo courtesy of Sony Corporation.*

8.7 VIDEO SERVERS

There are many applications for random-access storage of large amounts of video. For example, in large systems for nonlinear editing, it is desirable to centralize the storage of video material so that multiple editing workstations can have access to all the source material. This gives rise to the concept of a *video server*, a unit that stores large quantities of video with the capability for real time access by multiple users. Such a unit is shown as a peripheral to the Sony editor in Figure 8.11. Applications for even larger video servers occur in interactive video systems or *video on demand* (VOD) systems.

A video server generally consists of a number of hard disk drives connected to one or more high speed data buses and controlled by a PC-based system. The primary objective of a server's architecture is data throughput, since simultaneous users of video can multiply the combined data rate to a large value. A secondary objective of a server may be data reliability, since the existence of multiple disk drives offers the opportunity for redundancy.

8.7.1 RAID Systems

Architectures for combining multiple disk drives for redundancy and/or increased data throughput are called *redundant arrays of inexpensive disks* (RAID). Different RAID configurations offer varying degrees of redundancy and multiple access. Typical modes include the following:

- RAID 0 connects all the drives as one large storage unit, making available the entire capacity, but with no redundancy. Files written to a RAID 0 array are distributed to all the drives in a way that gives maximum throughput.
- RAID 1 mirrors the data between pairs of drives. Thus, the pairs of drives back up each other. Storage capacity is halved, but there is 100% backup in the case of drive failure.
- RAID 5 requires at least three drives and data are written across all three drives. However, error-correction files are also created and stored on all three drives, providing the capability to correct errors occurring on any of the drives. The generation of EC files slows down recording, but playing is still fast. Of course, the EC files take up some space, so storage capacity is reduced, but not as much as with RAID 1.

The operation of all the RAID modes is transparent once the system is set up. Backup or error correction occurs automatically at playback time.

8.8 CONCLUSION

Modern editing capabilities have made the video recorder into a tool for program creation that rivals the flexibility of motion picture film production, but with the further advantages of the immediacy and efficiency of electronic video recording. Editing equipment is available with full capabilities for all markets from professional program producers to home users.

REFERENCES

[1] SMPTE Standards. Society of Motion Picture and Television Engineers, 595 W. Hartsdale Ave., White Plains, NY 10607-1824. See also http://www.smpte.org

[2] Luther, A. C., *Principles of Digital Audio and Video*, Artech House, Inc., Norwood, MA, 1997.

9

Recorder System Features

Video recorders never exist completely by themselves—tapes must be loaded, signals must be connected, and the recorders must be controlled by their users. These requirements for use introduce many considerations into recorder design. These *system features* are covered in this chapter.

The extent of a recorder's system features depends on the size and sophistication of the system in which it resides. The recorder in a home camcorder is controlled by the user who is carrying it, and it may not have any external requirements except for loading tapes and connecting to a TV receiver for viewing. At the other end of the scale, a recorder in a postproduction editing suite has numerous signal interfaces, sophisticated remote control, and features that deal with all the different modes in which it may operate. This chapter covers that range of application.

9.1 PACKAGING

Recorders are packaged in at least five different ways, depending on their type of mounting. These are shown in Figure 9.1.

- *Tabletop mounting*—This is the configuration of the typical home VCR, where the "tabletop" is usually a TV set top. This type of mounting is also often used for small editing setups of two or three VCRs, monitors, and edit controllers. Tabletop VCRs are generally designed to be stackable as well.

- *Rack mounting*—The standard 19-in rack is widely used in fixed installations where a large amount of equipment is used together. The 19-in dimension refers to the distance between the two vertical equipment mounting flanges of the rack; rack heights can be any value up to about 7 ft. Units designed for rack mounting have flanges at either side of the front panel to mate with the mounting flanges of the rack. Often, tabletop units that are less than 19 in wide can be rack-mounted by adding mounting brackets to the sides of the unit.

Figure 9.1 *Recorder packaging: (a) tabletop, (b) rack-mounted, (c) camcorder, (d) field portable, and (e) dockable. Photos courtesy of Sony Corporation.*

- *Camcorder*—This is the familiar combination of a recorder with a camera. Camcorders are generally intended to be hand carried at least some of the time and must be light in weight. This results in the recorder being closely integrated with the camera both electrically and mechanically, and camera and recorder cannot be separated.

- *Field portable*—This is a standalone recorder in a portable package; it is intended to be cable-connected to other units in a field system. The requirement for light weight is the same as a camcorder's recorder but, in this case, the recorder has its own cabinet.

- *Dockable*—This is a variation of the field portable recorder that is intended to be coupled to cameras having a special mounting interface for a recorder. It provides the handling convenience of a camcorder while allowing camera and recorder units to be separated for individual use or for combination with other cameras or recorders. Dockable recorders are sometimes designed to be usable as field portable units, too.

Of course, there are many variations of these basic packaging concepts. Also, the application of the features discussed in the rest of this chapter depends on the intended use, as indicated by the packaging.

9.2 SYNCHRONIZATION

A recorder operating by itself has to synchronize only to the records it is playing and produce a signal that is usable by a display device. In this case, the recorder synchronization can make use of the ability of the displaying device to synchronize itself to a somewhat non-standard signal. That is the case, for example, with a home VCR that produces a signal with time base instability, which would be unacceptable in a system but works fine with TV receivers or displays because their synchronization circuits can follow the instabilities of the VCR signal and produce a stable picture.

However, when a recorder operates in a system with other recorders, signal sources, and video switching equipment, it must produce a stable, standard signal that is phase locked to the system synchronizing reference. That is necessary so signals can be switched, mixed, or otherwise processed together without any instabilities, interruptions, or timing distortions. It is a more demanding requirement than that of the home VCR and is one of the reasons why system recorders are more expensive.

Another synchronization feature that is becoming more common is the ability to synchronize to different video signal standards, such as NTSC and PAL. In the digital era, this will be even more common because DTV standards allow a range of scanning parameters to be used. All DTV equipment will be able to synchronize within this range.

Finally, for editing use, a recorder in playback should be able to synchronize to a specific time code stream, as explained in Section 8.3.1.4.

9.3 CONTROL

Recorders need controls for operation, setup, and testing. Each of these is explained below.

- *Operating controls*—These are controls for the starting, pausing, or stopping of playing or recording. All recording devices must have these controls, locally or remotely.
- *Setup controls*—These are controls for establishing the conditions needed for correct operation with the signals and system at hand. Such controls are minimized on home products, but they are important for professional products to provide the flexibility required by that market.
- *Test controls*—These are controls that facilitate testing the performance of systems. They exist only on professional products.

As an example of the controls provided on different classes of equipment, Table 9.1 lists the controls that may be provided on a home VCR, a professional VCR, and a DVD disc player.

Table 9.1

Controls for Various Recording Devices

Category	Home VCR	Professional VCR	DVD player
Operation	STOP	STOP	STOP
	PAUSE	PAUSE	PAUSE
	PLAY	PLAY	PLAY
	FAST FWD	FAST FWD	FAST FWD
	REWIND	REWIND	FAST REVERSE
	RECORD	RECORD	EJECT
	EJECT	EJECT	
		SHUTTLE-JOG	
		Editing controls	
Setup	Tape speed	Input selector	
	Input selector	A/V input levels	
	Channel selector	Audio output levels	
	Audio input levels	Insert/assemble	
	Programming menu	Channels to edit	
	Up-down buttons	Time code setup	
	Menu button	Remote/local	
	Select button		
Display	TV channel	Audio levels	Mode
	Tape timer	Modes	Play time
	Audio levels	Time code	
		Status	
Test		E-E selector	
		Color bars	
		Leader countdown	

9.3.1 Home VCR Controls

A design objective for home products is to simplify operation as much as possible. Even with our best efforts, many consumers still think a VCR is too complex for them, and the more advanced features are seldom used. The items listed in Table 9.1 are the bare minimum to operate a recorder for general playback and recording of programs from a TV receiver. The commands, STOP, PAUSE, PLAY, FAST FWD, REWIND, and EJECT, are needed for simply playing tapes, searching them for specific points, and rewinding. The remaining controls are needed for recording.

The "Programming Menu" control is provided for setting up delayed recording of a TV program; it is the most complex feature and also causes the most trouble with consumers not understanding how to use it. The basic requirements of delayed recording are to specify the TV channel or other source to be recorded and the times for starting and stopping of recording. Because of cost, most home VCRs do not provide separate controls for each of these requirements, but rather, they take the user through a series of steps that

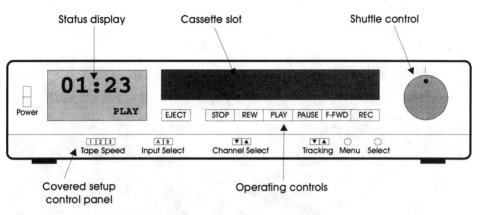

Status display   Cassette slot   Shuttle control

Power   PLAY   EJECT | STOP | REW | PLAY | PAUSE | F-FWD | REC

1 2 3   A B   ▼ ▲   ▼ ▲   ○   ○
◄ Tape Speed   Input Select   Channel Select   Tracking Menu Select

Covered setup   Operating controls
control panel

Figure 9.2 *Diagram of the control panel for a hypothetical deluxe home VCR.*

accomplish that setup of everything by using a single up-down button set, which may even be combined with another function, such as tracking. The display for the programming setup is usually provided by overlaying the TV screen with a menu display. Menu selections are highlighted on the screen, but the menu must be traversed by using the same up-down buttons in combination with a single "Select" button. The process becomes even more complex because features to review the programming or change it must also be provided. This process may be familiar to a computer person, but it is difficult for the average consumer. Such a complex approach is used because it is much less expensive to program the VCR's internal computer to provide the display and the step-by-step process than it is to provide dedicated controls for each of the functions required. Furthermore, consumers are also put off by complex-looking control panels because they think they will be difficult to learn to use.

Figure 9.2 is a drawing of the front panel for a high-end home VCR having the features described in this section.

9.3.2 Professional VCR Controls

Figure 9.3 shows the control panel of a professional VCR, the Sony DNW-A75. The categories of knobs and switches are called out in the figure and listed in the second column of Table 9.1.

The operating controls of a professional VCR are much the same as a home VCR except that there are usually advanced controls such as shuttle-jog that occur only on high-end home units. That control and other additional ones are used for implementing editing on the VCR, which is needed when a separate edit controller is not available.

Professional VCRs generally have more audio channels than home units, so there are more selector switches for choosing record sources and edit modes, and more knobs for setting audio levels. Video input level controls are usually also provided, and output level controls may also be included. Another area that adds a number of controls is the time

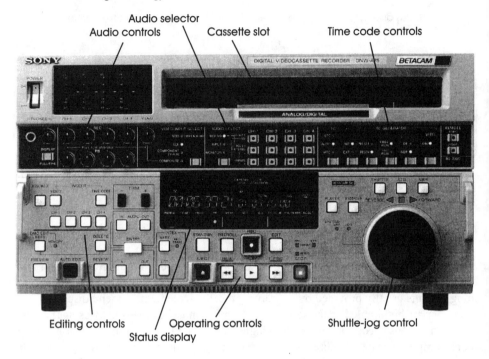

Figure 9.3 *Control panel for the Sony DNW-A75 Digital VCR. Photo courtesy of Sony Corporation.*

code capability, which requires choice of synchronization modes and selection for linear time code and/or VITC modes. Professional units also always have remote control features so they can be connected into routing or editing systems, and selectors are needed to choose remote modes.

Compared with home VCRs, professional units lack the preprogrammed delayed-recording feature, which has no use in a professional environment. Thus, the programming controls do not exist for that feature. This is made up for by the editing features, which are unusual on home recorders, except for camcorders. In that case, assemble editing is almost always provided because it allows a tape to be produced from multiple shots without interruptions between the segments. Many camcorders have an "auto-assemble" feature, where suitable pre-rolling is provided to make good edits every time the record mode is started.

Professional recorders generally do not shy away from complex control panels, and professional users are not put off by complex controls; in fact, they are comfortable with them and view the fancy control panel to mean the equipment has more features. Other reasons for many controls is to provide instant access to any one of them, and the position of switches shows the equipment status at a glance.

9.3.3 Computer-Based Controls

In this era of computers everywhere, recorder controls do not have to be actual knobs and switches—they may be software controls that exist only on a display screen, as was described above for programming a home VCR for delayed recording. This can be extended as the basis for all the controls.

Essentially all recorders use a computer with software for their internal control systems. Thus, if a suitable display exists with user-input controls, almost any type of control panel can be easily programmed on the computer. This offers a tremendous degree of flexibility—control panels can be customized for the particular application of the recorder or even for different users of the same product. (See also Section 9.7.2.)

9.4 REMOTE CONTROL

The purpose of remote control features is twofold: (1) to allow manual control from a remote location, such as a studio control room, or across a living room in the home environment; or (2) to allow control from a remote computer, such as an edit controller. Remote control can provide the same facilities that are available locally on the recorder although, in some cases, certain local features may not be remoted.

Remote control is generally digital and is transmitted according to one of a number of standard protocols. As with all communications systems, there are several levels of protocol, ranging from the physical system of wires, radio-frequency links, or fiber-optic connections, to the various levels of data protocols needed to complete a system. These are discussed below.

9.4.1 SMPTE 275M

The SMPTE and the European Broadcasting Union (EBU) jointly developed a network standard for control and data use in small to moderate-sized facilities for television and audio program production, postproduction, or distribution. This is known as *ESlan-1*, which is based on the widely used ANSI/IEEE 802.3 (Ethernet) physical hardware for computer networks. It is defined in accordance with the OSI protocol layers (see Section 1.8.3). The governing standard for the physical, data link, network, transport, and session layers is SMPTE 275M, but there are many other documents referenced by this standard.

The network layer uses the *Internet Protocol* (IP), which is another widely used standard. Each unit on the network is given an IP address, which is used for communication with any other unit. The actual control information is transmitted by data packets in accordance with other SMPTE standards that define control data fields.

9.4.2 RS-232C and RS-422

RS-232C is a standard published by the *Electronic Industries Association* (EIA) for serial interfacing between computers and peripheral devices such as modems or printers. It is a hardware standard for serial asynchronous data transfer using 9-pin or 25-pin connectors.

Only two wires are used for data (one in each direction); the remaining wires are for control. A *handshaking* protocol is used to provide *full-duplex* (simultaneous bidirectional) data transmission over a single cable at rates up to 115 kbps. Handshaking means that the source and receiver exchange control information before and after each data packet transmission. In RS-232C, a data packet is 1 byte.

RS-232C is widely used for connecting recorders to editing systems. This protocol provides for raw data transmission; it says nothing about the format of that data. Thus, a further standard is needed to define the data format for remote control information. RS-232C normally operates at data rates of 20 kbps, although it can be pushed up to 100 kbps over short distances. The recommended maximum distance for RS-232C communication is 15 m.

RS-422 is a higher performance serial communications standard that is capable of greater distances and greater data rates (up to 10 Mbps and 1,200 m) than RS-232C. It is generally required for large systems that involve distances between units greater than 15 m, as well as a greater number of communicating units.

9.4.3 Infrared Remote

The living room environment in the home is ideal for remote control using infrared (IR) transmission; this is proven by its popularity for control of TVs, VCRs, disc players, etc. Infrared has not been much used in professional circles, however, which is probably because of its limited range and reliability in complex reflective environments such as in TV studios and editing rooms.

IR remote controls in the home generally follow a format that allows remotes to be interchanged between units by reprogramming them. Some units are automatically programmable simply by entering the proper device code, which then allows the remote to access control parameters in its memory for the device identified by the code.

9.5 VIDEO INTERFACES

The analog video world has standardized on 75-ohm coaxial cable for video signal interfacing. This is inexpensive and reliable for distances up to 1,000 ft or more, if suitable equalization is used. In the professional video world the BNC connector is standard, but in the home it is the RCA "phono" connector; these are shown in Figure 9.4. It is simply a cost-performance trade-off—BNC connectors have a bayonet locking mechanism, while the RCA phono connectors just have spring contacts.

As with the cables, analog video signals are also well standardized. For NTSC signals, the standard is SMPTE 170M, which defines video waveforms for 525-line systems for studio applications.

Digital standards are, of course, much newer, but a pattern has already emerged. For 525- and 625-line systems, the scanning and sampling standard is ITU-R Rec. BT.601-5, which provides a range of choices for component digital systems (see Section 1.6.2.1). Most other standards for communication or compression of 525- and 625-line signals

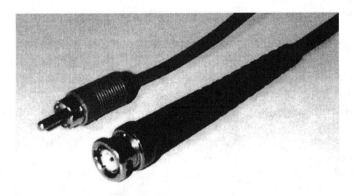

Figure 9.4 *Video connectors: RCA phono (left) and BNC.*

specify Rec. BT.601-5 as their input format. Thus, digital interface standards have two or more levels—the first being the scanning standard, such as Rec. BT.601-5, and the second being a communication standard, such as SMPTE 259M or IEEE 1394. A third level is interposed when compression is used.

Digital interfaces are classified according to how many data lines they use. When all data passes over a single wire, the interface is *serial*, meaning all bits are transmitted in sequence through the interface. If video is transmitted serially as a two-level binary signal (see Section 2.2.7.2), the bandwidth for uncompressed signals can become very high. This can be reduced by using multilevel modulation, which requires a linear transfer characteristic. Recording channels are nonlinear, so multilevel modulation is not applicable, but it can be used for cable transmissions. The 8-VSB modulation used in the ATSC DTV standards is a possibility, but so far, this has not been used for anything except broadcasting (see Section 1.8.3).

Digital bandwidth can also be reduced by using *parallel* transmission, where the signal is distributed to a multiplicity of wires. Standards exist for this (SMPTE 125M, 244M), but parallel cables are expensive and limited in length because of the need for precise matching of delays between the parallel wires. They are seldom used except for interfacing between units within racks. Serial interfacing is preferred for most connections between units outside of racks. With the development of dedicated ICs for serial processing, serial links have become quite economical. The rest of this section discusses some serial interface standards used by recorders.

9.5.1 SMPTE 259M

SMPTE standard 259M, also known as the *serial digital interface* (SDI), defines a serial interface for uncompressed 525- and 625-line video signals at data rates up to 360 Mbps. It is a hardware and data format standard that uses 75-ohm coaxial cables with BNC connectors—the same as used for analog signals in professional video systems. SMPTE 259M is described in Section 1.6.2.3.

9.5.2 IEEE 1394

SDI is a *point-to-point* interface, servicing one source and one receiver connected by one cable. A more flexible architecture is the IEEE 1394 standard that uses a bus structure; multiple units are connected by cables in a loop-through configuration. Any two units on the cable loop can communicate simply by exercising a protocol to acquire some of the bus capacity for the duration of their communication. 1394 is a general-purpose data bus that does not care about the content of the data it transmits, so it can handle video, audio, compressed or noncompressed, or computer data at any time. This is described further in Section 1.6.2.4.

The 1394 interface was developed for interconnection of VCRs and other digital units in the home. However, it is capable of much more than that and will surely see wide application for both video and computer devices in both home and professional environments.

9.5.2.1 MPEG-2 Transport Stream

The MPEG-2 standard for compression (see Section 1.6.2.6) includes a packetized stream data format that is widely applicable, even when compression is not used. This can be used with any general transmission standard such as IEEE 1394. The packetizing method of MPEG-2 was shown in Figure 1.8.

Recorders using compression generally invoke the MPEG-2 transport stream. This is true in the DV (see Section 10.2.2), Betacam SX (see Section 10.2.6), and DVCPro (see Section 10.2.3) formats. The transport stream is also used by the ATSC DTV standards.

9.6 AUDIO INTERFACES

Analog audio signals are handled on coaxial cable or sometimes twisted-pair cables. Professional audio uses the XLR connector series, which provides balanced lines (two signal wires), and it is more immune to interference and noise over long distances. Home audio uses the RCA phono connectors; these are unbalanced and have limited distance capability and noise immunity. Analog audio signals are baseband audio without any encoding.

Digital audio systems use the same connector types in professional and home environments, but, of course, signals are digitally encoded. There are several different standards, including the Dolby Digital standard (see Section 7.5.3.2) and the MPEG-2 audio standard (see Section 7.5.3.3).

9.7 USER INTERFACES

The design of user interfaces for electronic products such as recorders or computers is a complex subject that has spawned its own books [1]. This section does not cover that entire field; it just discusses some of the user-interface issues that are unique to recorders.

9.7.1 Shuttle-Jog Control

Recorders have the unique requirement of moving back and forth in a recording to view different parts of it or to locate a specific frame for editing. To do this, it must be able to play a recording at variable speeds from very fast to very slow compared with normal speed. This is called *shuttling*. Recorders, of course, must have the mechanical and electronic capabilities to do this (it adds to the cost, so not all recorders have variable-speed play capabilities), but then there also must be a means for the user to control the recorder in this mode.

Shuttling can be accomplished in a crude way on tape decks with the fast forward and rewind controls found on every system. However, this proves very awkward because of the mechanical response of the tape deck, which is not instantaneous and has a significant inertia effect. Much better control is obtained with a shuttle knob that controls play speed smoothly from full rewind speed through zero speed to full forward speed. The inertia of the tape deck still causes a delayed response, but the user can become aware of that and suitably anticipate his or her actions.

The shuttle knob still is difficult when the play position needs to be moved by one or at most a few frames. This is known as *jogging* the recording and is better done with a pair of buttons that move the play position by one frame each time a button is pressed. Various designs have been made to combine these two functions into the same control, such as by having a shuttle knob that switches to jog operation if it is pushed in. In that position, the playing stops and the knob becomes a jog control that moves one frame for each specified amount of rotation (20°) in either direction. When the jog mode is released by letting the knob come back out, it switches back to shuttling but it sits at the zero-speed position.

Shuttle and jog control is even more effective with disk recorders because there is essentially no delay or inertia in the response. However, the user control requirements are much the same and the shuttle knob and jog controls or the combined control are still the best way to give control to the user.

9.7.2 Computer-Style Interfaces

A *graphical user interface* (GUI), as used on PCs, consists of a pointing device (mouse) and a graphical display. This is suitable for applications that have a large number of controls or that can benefit from the flexibility of presentation available with the graphical display. So far, GUI interfaces have not appeared on individual recorders but they are used in editing systems everywhere except for the simplest units.

Dedicated control panels with separate switches, buttons, or other controls for each function, have the advantage that any function is instantly available by directly operating the appropriate controls. However, for complex devices such as recorders with editing, dedicated control panels are expensive and may be physically large. Replacing these with a CPU, display, and pointing device may save cost or reduce size, and also can add as much flexibility in the programming as the designer wishes to provide. A disadvantage is that the user may have to go through several steps to access a specific control, which may be unsuitable when fast response is required. This problem can be minimized by carefully designing the screens to assure that controls that are used together in a particular mode of operation all exist on the same screen.

A particularly interesting type of GUI is the *touch-screen display*, which combines the display and pointing device functions into one unit. However, these tend to be expensive and have not been used much on recorders or editing equipment.

9.7.3 Cassette Loading

Mechanisms for loading the tape from a cassette into a tape deck were discussed in Section 5.7. From the user interface point of view, cassette loading is simple—one either inserts the cassette into a slot or opens a door and puts it in—those seem to be the only choices. Of course, recognizing that users are most ingenious in finding ways to do things wrong, the interface must deal with all the ways a user may insert the cassette incorrectly. Such possibilities should result in rejection of the cassette and should not damage the cassette or mechanism.

The type of cassette loading affects the configuration and range of application of the recorder; this occurs because the cassette size and tape path considerations affect the layout and size of the tape deck. Decks are usually relatively thin in a direction perpendicular to the cassette body and have extended dimensions in a plane parallel to the cassette body. That means a cassette-loading slot appears on an edge of the tape deck, or a cassette-loading door is on top of the tape deck. Applications such as rack mounting or desktop mounting with stacked recorders favor slot loading. A camcorder, however, usually needs to have its deck vertically oriented to minimize camcorder width, which allows for a loading door on the side of the unit with the cassette inserted from the top. This is very convenient because cassettes can be changed without removing the camcorder from the operator's shoulder.

9.7.4 Adjusting Analog Parameters

The most common analog control is the rotating knob. However, in digital systems, there are additional opportunities for adjusting "analog" parameters. Such parameters are, of course, digitized in a digital system, but they have enough quantizing levels that they can be considered analog or continuous. That usually means they are digitized with 8 or more bits.

The most common "digital" control for analog parameters is the "up-down" control. This consists of two buttons that cause the function value to increase or decrease when pressed. Usually, a single button press will change the digital value by one quantizing level; holding the button down longer than a certain minimum time will cause the value to change steadily in the chosen direction. A single set of up-down buttons can control a number of different parameters by providing a selector switch to choose the parameter to be controlled by the buttons. Of course, the control system in the device must have memory to remember the setting for each parameter as the selector switch is moved. If it is necessary for the user to see the value being adjusted by the up-down control, a numerical display can be provided.

In a GUI digital interface, there are more possibilities for analog controls. A rotating knob itself can be modeled, or a slider control can be provided that shows the current value by the position of the slider. Of course, up-down controls can also be implemented in a GUI. It is even possible in a GUI to adjust two-dimensional analog parameters by displaying a rectangular array where each position in the array represents a particular combination of the two parameters. This is often used in PCs for setting of colors. Of course, color is a three-dimensional value, so the array can only display two dimensions at a time and another control must be used to set the third value; still, it is a very convenient way to set and display color selections.

9.7.5 Editing Controls

The controls needed for editing were discussed in Chapter 8; Section 8.6 showed some specific designs for edit controllers.

9.8 CONCLUSION

All recorders are used with other equipment in a system. Even a home VCR or camcorder must interface to a TV receiver or other display. Professional recorders integrate into studio or editing systems and must have many features to interface to the rest of the system, including synchronization, signal connections, and remote control. The design of these features and the standards chosen are a large factor in the ultimate success and competitiveness of a recorder.

REFERENCE

[1] Laurel, Brenda (ed.), *The Art of Human-Computer Interface Design*, Addison-Wesley, Reading, MA, 1990.

10

Tape Recorder Products

Recorders are constrained by the medium and format that they use. Within those constraints, they may be further categorized by their configuration and intended use. Mature recording formats offer a wide range of configurations suited to many different use scenarios.

In a production-postproduction environment, program content moves between production and postproduction via recorded media; this type of operation cannot be supported unless the format offers equipment for both areas—camcorders and portable equipment for production, and editing recorders for postproduction.

In the event that all equipment types are not available in the same format, then a format conversion must occur at some point. This is an undesirable extra step. In at least one case, however, the extra step is worth the trouble—that is for nonlinear editing using disk recording. At present, there is no disk recorder suitable for production use. Production is done on tape and converted to disk when the media reach the editing site. The advantages of nonlinear editing are so great that most people accept this extra step.

This chapter covers both analog and digital recording equipment, even though digital recorders represent nearly all new product announcements at this time. Nevertheless, millions of analog recording units are in use and will continue to be used for many years. Thus, it is not yet appropriate to ignore analog recording equipment. No attempt is made to cover all the formats and equipment in the market; systems were chosen for description here solely to provide a diversity of approaches that will show the range of recording equipment currently available.

The recording field is moving rapidly in the development of new digital formats and equipment as well as finding new markets and uses. Any description of actual equipment, as is presented in this chapter, must necessarily become rapidly out of date as the field progresses. The general principles of equipment design as displayed by the units described here, however, remain relevant even as new equipment evolves.

10.1 ANALOG RECORDERS

At this time, digital recorders are increasingly dominating the professional market. In the home market, however, although digital recorders have been introduced, they are still expensive for that market and make up only a small percentage of new purchases. Thus, this section covers only one analog professional system (SMPTE Type C), but two major analog formats for the home market are covered—VHS and 8-mm. These latter two systems both have enhanced versions that provide somewhat higher picture quality for high-end home and semi-professional markets. These are called S-VHS and Hi-8.

Another analog professional system that is widely used is Sony's Betacam SP system. This is a 1/2-in tape system with tape decks and record formats similar to the newer Betacam SX digital recorders. Betacam SX is covered in Section 10.2.6.

10.1.1 SMPTE Type C

In the early 1970s, Sony developed a high quality helical scan recording system using 1-in tape that soon became the leading recorder for the TV broadcast business. This replaced the quadruplex transverse-scan system then in use and was standardized by the SMPTE as the *Type C* recording system. Type C recorders are still widely used in broadcasting, although they are no longer being manufactured.

The Type C format is intended for reel-to-reel tapes, not cassettes. Tape loading requires manual threading, although some decks provide some assistance with that process. In any case, a recorded tape is handled outside of the machine as a single reel, with a loose end of tape extended for threading. In one respect, this is better than a cassette because it is smaller than a two-reel cassette would be, but it is worse for protecting the tape, since the end of the tape must always be accessible. It is also awkward to remove a tape from a deck without rewinding, because then two reels must be removed but kept together somehow with a loop of tape between them. The potential for tape damage is much higher than with cassettes.

10.1.1.1 Type C Helical Scanner

The Type C scanner is a near-360° omega-wrap configuration (see Section 4.2), which is sometimes called a "one-head" video recorder. The wrap angle on the scanner is large enough to record an entire active picture for one field in one helical track on each rotation of the scanner. However, a second "sync" channel is provided to record VBI information in short tracks at the edge of the tape, so the entire signal is actually recorded. This would not be necessary if the VBI only contained sync information because it could be regenerated on playback, but it is valuable when the VBI also contains data or other signals, which is becoming increasingly common.

The head layout of the Type C helical scanner is shown in Figure 10.1; there are separate record/playback, erase, and confidence playback heads for both video and sync channels, giving a total of six heads on the drum. Confidence playback heads allow replay during recording, which is a significant feature in professional recording and editing. The

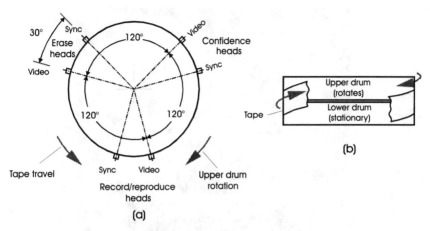

Figure 10.1 SMPTE Type C head layout on helical scanner: (a) top and (b) side view.

record/replay head for the main video is mounted on a piezoelectric bimorph structure that allows transverse movement of the head for automatic tracking control and variable-speed playback without picture disturbances (see Section 5.5.5). The upper drum rotates at 59.94 rps for NTSC or 50 rps for PAL and SECAM, and the lower drum is fixed for edge guiding the tape around the scanner. As in all helical scanners, the scanning drum is servo controlled for synchronism with the video signal system.

10.1.1.2 Type C Tape Deck

Figure 10.2 is a photograph of a Type C tape deck, the Sony BVH-3000. The scanner is centrally located with the reels above, supply on the left, and takeup on the right, and the control panel is at the bottom. This gives the best access for manual threading. The distance between the entrance and exit tape guides on a Type C scanner is too close for convenient threading, so they are designed to move out of the way in a special "Load" mode. For normal running of the deck, the guides are moved together by a motor drive and mate against mechanical stops to achieve the necessary precision for helical scanning. Other tape path elements are conventional: tension arms, capstan and pinch roller, longitudinal head stacks, full-width erase head, and so on. The Sony recorder in the figure also has an air-assisted threading feature where the operator simply feeds the end of the tape along a slot while compressed air directs it into the tape path and around the takeup reel.

The Type C package shown is called a *tabletop* configuration; it has handles for transportability, and it also can be equipped with hardware for rack mounting.

10.1.1.3 Type C Track Records

The track records for Type C are specified in ANSI/SMPTE Standard 19M-1996 and are shown in Figure 10.3. The main helical tracks cover most of the tape width, with the sync

Figure 10.2 The Sony BVH-3000 SMPTE Type C recorder. Photo courtesy of Sony Corporation.

head using a small area near the reference edge of the tape. At the far edge of the tape are the two longitudinal program audio tracks, and a third audio track for cueing or auxiliary use is along the reference edge. The control track is positioned between the spaces for the video and sync helical tracks.

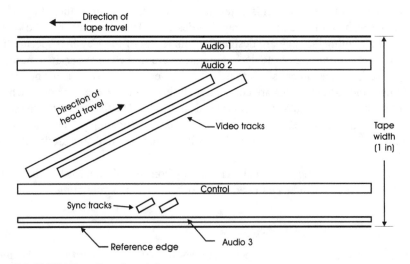

Figure 10.3 SMPTE Type C track records.

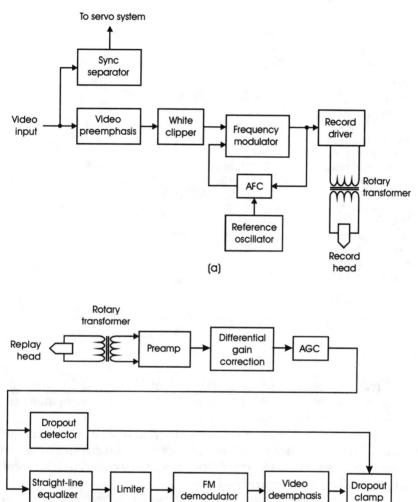

Figure 10.4 SMPTE Type C (a) record and (b) replay signal processing.

Table 10.1
Type C Video FM Carrier Frequencies

	Carrier frequency (MHz)	
Signal level	NTSC	PAL
White	10.0	9.3
Blanking	7.9	7.8
Sync tip	7.06	7.06

Table 10.2

Type C Typical Specifications

Item	Specification
Tape width	1 in (25.4 mm)
Tape speed	244.0 mm/s
Recording time	126 min with 11¾-in reels
Scanner diameter	134.6 mm
Scanner no. of heads	6 (2 record/play, 2 erase, 2 confidence playback)
Video writing speed	25.4 m/s
Video track width	0.125 mm
Video signal standards	NTSC, PAL, or SECAM composite analog video
Video bandwidth	NTSC: 4.2 MHz; PAL/SECAM: 5.0 MHz
Video SNR	NTSC: 49 dB; PAL/SECAM: 45 dB
Video linearity*	DG—< 4%, DP—< 4°
Video transient response	< 1% K-factor
No. of audio channels	2 program (Pgm), 1 cue/time code
Pgm audio bandwidth	50 Hz to 15 kHz
Pgm audio SNR	56 dB
Pgm audio distortion	< 1%
Audio wow and flutter	< 0.1% rms
Equipment weight	70 kg

* In terms of differential gain (DG) and differential phase (DP). Applies only to NTSC and PAL.

10.1.1.4 Type C Signal Processing

The Type C recorders use frequency modulation (FM) of the composite video signal, with carrier frequencies as shown in Table 10.1. Record and playback video signal processing is block diagrammed in Figure 10.4.

For recording, sync information is separated for servo control and the video is preemphasized, white peaks are clipped, and the signal is FM-modulated for recording. The FM frequencies are stabilized by an automatic frequency control (AFC) system that uses a precise oscillator for a frequency reference.

On replay, the head output is amplified and processed to correct differential gain errors. The signal level is controlled by an automatic gain control (AGC) circuit; it is equalized and limited before FM demodulation. Following that, the video is deemphasized, and dropout errors are suppressed (clamped) by a signal from a dropout detector that senses when the FM signal level from the AGC momentarily drops below an acceptable level; otherwise, the FM demodulator could produce a burst of noise that might overload subsequent circuits. Finally, the output video goes to a time base corrector (TBC) for stabilization of time base errors.

10.1.1.5 Type C Specifications

Table 10.2 lists some of the characteristics and performance of a typical Type C recorder. These recorders are physically large, obviously because of the large tape and reel sizes.

First-generation video performance is excellent but, being analog, these systems suffer from accumulated degradations on multiple generations. Acceptable video for most purposes, however, is usually produced through three or four generations.

10.1.2 VHS and S-VHS

Home VCRs were introduced in the late 1970s for consumer recording of off-air TV programs for later replay. The market really exploded, however, with the availability of prerecorded tapes of film movies, music videos, TV programs, and other materials. The VHS (Video Home System) format using 1/2-in tape is the leading format in this market.

The introduction of camcorders also expanded the home VCR market, allowing consumers to record video of family events, travel, and other scenes. Although VHS camcorders were introduced, they were physically large for consumer use, and generally required mounting the camcorder on the user's shoulder. Smaller formats were designed for camcorder application, including the 8-mm format based on smaller tape (8-mm), and a small-cassette version of VHS (VHS-C), that is compatible with full-size VHS.

VHS and 8-mm formats provide a performance level that compares to "average" TV reception, which is not as good as the best that broadcast TV can do. However, it has been accepted by consumers for the uses described above. VHS and 8-mm equipment is inexpensive and has seen use in market areas outside of the home, especially the semi-professional markets for program producers who do weddings and other special events for consumer viewing. However, performance is marginal for the multigeneration operation required to edit and assemble a program. This has led to higher-performance versions of both VHS and 8-mm systems that retain a lot of equipment commonality but use more expensive tapes and faster tape speeds to achieve better pictures through multigenerations. These higher-performing systems are the Super-VHS (S-VHS) and Hi-8 formats.

10.1.2.1 VHS Helical Scanners and Tape Decks

The VHS system is standardized as SMPTE Type H, described in ANSI/SMPTE Standard 32M-1993. The format calls for 1/2-in tape and a two-head, 180° omega-wrap scanner (see Section 4.2), but compatibility can also be achieved with a four-head scanner having 270° wrap. This has the advantage that the scanner diameter is smaller and it is used especially in camcorder applications.

VHS decks have been produced in tabletop, rack-mounted, and camcorder configurations. Some of these were described in Chapter 5.

10.1.2.2 VHS Records

Since there are at least two heads writing the video tracks, VHS can enjoy the advantages of azimuth recording (see Section 3.2.1.1). VHS provides three tape speed options for extending the play time at the expense of picture and sound quality; these are Standard

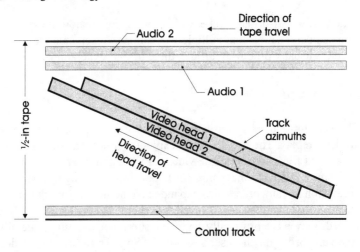

Figure 10.5 VHS track patterns.

Play (SP); Long Play (LP), which cuts tape speed in half for twice the playing time; and Extended Play (EP), which reduces tape speed to 1/3 of SP for 3× playing time.

The effect of lowering tape speed is poorer video SNR and poorer audio bandwidth and SNR, which is bad enough that the lower tape speeds are seldom used by anyone conscious of picture and sound quality. Because of this, high-end VHS recorders offer an alternative audio system that uses two FM subcarriers at 1.3 and 1.7 MHz helically recorded under the video in the same tape area as the video tracks. This is referred to as VHS-Hi-Fi and gives better audio performance that is independent of tape speed.

The track patterns are shown in Figure 10.5. The azimuth-recorded video tracks are shown, with two longitudinal tracks for stereo audio and one longitudinal control track.

10.1.2.3 VHS Signal Processing

Because of its extremely low video writing speed, and therefore low recording bandwidth, VHS uses the color-under system that was described in Section 7.3.3. The record processing for this system is shown in Figure 10.6, and FM carrier frequencies are listed in Table 10.3.

Video input is split into luminance and chrominance components by suitable filters. The luminance is preemphasized, excess amplitude peaks are clipped, and the signal is frequency modulated. The chrominance signal is down-converted to a carrier frequency equal to 40 times the horizontal scanning frequency and combined with the luminance FM for recording. Additional chrominance processing doubles the amplitude of the color burst to improve color synchronization on replay, and the down-converting carrier is phase shifted in 90° steps from line to line to reduce interference between adjacent tracks on replay. This is needed because the recorded wavelength of the chrominance signal is too long for the azimuth cancellation to be completely effective.

Table 10.3
VHS and S-VHS FM Carrier Frequencies (NTSC)

| | Carrier frequency (MHz) | |
Signal level	VHS	S-VHS
White	4.4	7.0
Blanking	*	*
Sync tip	3.4	5.4

* Not specified.

10.1.2.4 Features of S-VHS

The S-VHS format was designed to improve the performance of the VHS system for high-end consumer and semi-professional markets. This is accomplished by using higher-coercivity tape and higher FM carrier frequencies. This allows S-VHS luminance bandwidth and FM frequency deviation to be increased, although chrominance bandwidth remains the same as VHS.

S-VHS recorders also offer a special multiconductor signal interface that transmits luminance and chrominance on separate wires, called *S-video*, which improves performance in multigenerations because cascading of the chroma-luminance combination and separation processes is avoided. For the same reason, S-video also provides higher picture quality when connecting a recorder to a TV display that has a S-video input.

In principle, the S-video interface can be provided on any color-under recorder (not just S-VHS), but it is slightly more costly and, thus, may not be available on the lowest-cost units.

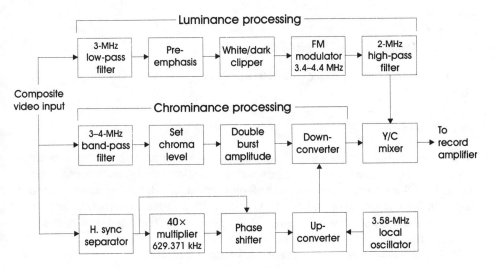

Figure 10.6 *VHS recording signal processing.*

Table 10.4
VHS and S-VHS Typical Specifications (NTSC)

Item	VHS	Specifications (both)	S-VHS
Tape width		1/2 in (12.65 mm)	
Tape speed		SP: 33.3 mm/s	
		LP: 16.7 mm/s	
	EP: 11.1 mm/s		
Recording time	2/4/6 hr*		2/4 hr*
Scanner diameter		2.44 in (2 heads) or 1.63 in (4 heads)	
Scanner no. of heads		2 or 4	
Video writing speed		5.8 m/s	
Video track width	0.058/0.029/0.019 mm*		0.058/0.029*
Horizontal resolution	240 TVL		400 TVL
Chrominance bandwidth		0.5 MHz	
Video SNR (SP mode)	45 dB		45 dB
No. of audio channels		2	
Audio bandwidth	50 Hz to 10 kHz		20 Hz to 20 kHz†
Audio SNR (SP mode)	45 dB		
Audio dynamic range			90 dB†
Equipment weight††		4.5 kg	

* A/B/C means SP/LP/EP.
† Hi-Fi digital audio mode.
†† Approximate for full-size VHS set-top configuration. VHS-C camcorders can weigh less than 1 kg.

Some S-VHS recorders also offer a digital audio system that provides 16-bit digital audio performance in two channels that are multiplexed onto the helical tracks along with the video.

10.1.2.5 VHS and S-VHS Specifications

Specifications and performance parameters for both VHS and S-VHS are given in Table 10.4.

10.1.3 8-mm and Hi-8

As the home camcorder market evolved, there was an opportunity to develop a recording system based on tape smaller than 1/2 in, which would allow camcorders to become smaller and lighter. Sony addressed this with the 8-mm format, which has become an important player in the home camcorder market. As with VHS, a further opportunity appeared to create an enhanced-performance version of this for high-end home and semi-professional markets. Thus the Hi-8 format was born.

8-mm and Hi-8 also use the color-under system for recording.

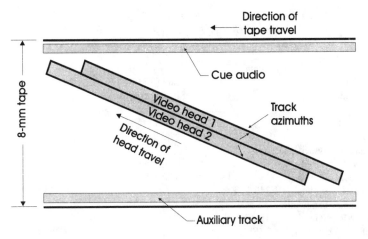

Figure 10.7 *Track patterns for the 8-mm system.*

10.1.3.1 8-mm Helical Scanner and Tape Decks

The 8-mm system uses a smaller scanner diameter than VHS, but it still uses the 180° omega-wrap configuration with two recording heads. Tape speeds and track widths are smaller to achieve competitive playing time in a smaller cassette. Signal performance is maintained by using improved tape and head technologies. As expected, the smaller scanner and cassette sizes allow tape decks to be somewhat smaller and lighter than VHS-C units.

10.1.3.2 8-mm Track Patterns

Track patterns for the 8-mm system are shown in Figure 10.7. This system recognizes that program audio performance of longitudinal tracks is marginal at best with the tape speeds and track widths available in these recorders. Thus, no longitudinal program audio tracks are provided. Two low-performance tracks provide for cue, time code, or auxiliary audio uses. Also, there is no control track. Tracking servo information is taken from the video tracks themselves. The standard includes FM audio recorded along with the video, and additional video track length is provided for the use of digital audio in the video tracks as an option. Thus, the total scanner wrap angle is 221°.

Table 10.5
8-mm and Hi-8 FM Carrier Frequencies (NTSC)

	Carrier frequency (MHz)	
Signal level	*8-mm*	*Hi-8*
White	5.4	7.7
Sync tip	4.2	5.7
Audio	1.5	1.5

Table 10.6
8-mm and Hi-8 Typical Specifications (NTSC)

	Specifications		
Item	*8-mm*	*(both)*	*Hi-8*
Tape width		8-mm (0.316 in)	
Tape speed		SP: 14.3 mm/s	
		LP: 7.15 mm/s	
Recording time		2/4 hr*	
Scanner diameter		39.9 mm	
Scanner no. of video heads		2	
Video writing speed		3.8 m/s	
Video track width		0.02/0.01 mm*	
Horizontal resolution	240 TVL		400 TVL
Chrominance bandwidth	0.5 MHz		0.7 MHz
Video SNR (SP mode)	45 dB		45 dB
No. of audio channels	2 (FM)		2 (Digital)
Audio bandwidth	50 Hz to 10 kHz		20 Hz to 20 kHz†
Audio SNR (SP mode)	45 dB		
Audio dynamic range			90 dB†
Equipment weight††		3 kg	

*   A/B means SP/LP.
†   Digital audio mode.
†† Approximate for full-size 8-mm set-top configuration. Camcorders can weigh as little as 0.5 kg.

10.1.3.3 8-mm Signal Processing

Signal processing in the 8-mm system is similar to that shown in Figure 10.6 for VHS except that FM audio is always included. FM carrier frequencies for 8-mm are shown in Table 10.5. Since the frequencies are slightly higher than VHS and the video writing speed is slower, the minimum recorded wavelength is significantly shorter, calling for more advanced head and tape technology.

10.1.3.4 Features of Hi-8

The Hi-8 system uses higher FM carrier frequencies (shown in Table 10.5) as in S-VHS to achieve greater video bandwidth. This requires further advances in head and tape technology, which has been achieved in these products. Hi-8 recorders generally also implement the digital audio option for improved audio performance.

10.1.3.5 8-mm and Hi-8 Specifications

Specifications for the 8-mm systems are shown in Table 10.6.

10.2 DIGITAL RECORDERS

Although the majority of video recorders now in use are analog, essentially all new product announcements in the last several years have been digital formats. This is expected to continue because the advantages of digital recording are overwhelming in all markets, and the high-cost image of digital technology is rapidly being laid to rest. Thus, the rest of this chapter will cover some of the digital formats that are enjoying high interest in the markets today.

10.2.1 SMPTE D-Series

Because digital recorders were at first expensive, it is only natural that they would initially appear in the broadcast and professional markets. These formats have been standardized by the SMPTE in a series beginning with D-1 and continuing today with D-7 and beyond.

10.2.1.1 D-1 Component Recorder

The first recorder of the D-Series, the D-1, is a component system based on the ITU-R Rec. BT.601-5 4:2:2 sampling structure, without compression. This is consistent with delivering the highest possible performance for SDTV program production and postproduction applications. Figure 10.8 is a photograph of a typical D-1 recorder in a tabletop configuration, the Sony DVR-2000.

Figure 10.8 *The Sony DVR-2000 D-1 component digital recorder. Photo courtesy of Sony Corporation.*

The D-1 format is defined by the following ANSI/SMPTE standards:

225M	Magnetic Tape
226M	19-mm Tape Cassettes
227M	Helical Data and Control Records
228M	Time and Control Code and Cue Records

Tape and Cassette—D-1 uses 19-mm (3/4-in) tape in cassettes that are available in three sizes. The tape deck automatically adjusts its reel drives and threading components to the size of the cassette during insertion. The maximum recording times for the three cassette sizes are: small cassette (S)—6 min, medium cassette (M)—34 min, and large cassette (L)—94 min.

Track Patterns—Figure 10.9 shows the track patterns for D-1. The helical tracks hold both video and audio and are recorded with perpendicular azimuth and guard bands. Three longitudinal tracks provide for control, time code, and cue audio recording. Four audio channels are assigned to data blocks in the center of each helical track, dividing the video content of each track into two sectors.

The helical tracks are recorded with four heads, with the data distributed between the heads so that the best possible concealment is available during a dropout or even when one head fails completely. The scanner rotates at 150 rps, and four tracks are recorded on each rotation (four heads), giving 600 tracks/s. In NTSC, there are thus 10 tracks per video field; in PAL there are 12 tracks per field. The data distribution strategy is designed so that a head failure will not cause complete loss of data at any picture position. This is a common technique in all multiheaded digital video recorders.

Signal Processing—A block diagram of D-1 signal processing is shown in Figure 10.10. Video data undergoes steps of encoding, shuffling, and outer error coding before being multiplexed with audio data that has been through similar processing steps. The combined data then receives inner error coding and randomizing, and is multiplexed with sync and ID information. Finally, the channel coding is applied and the data is distributed to the head drivers for recording.

Specifications—Physical and performance specifications for a D-1 recorder are shown in Table 10.7.

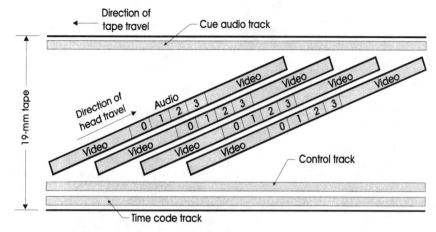

Figure 10.9 D-1 track patterns. Reproduced with permission from [1].

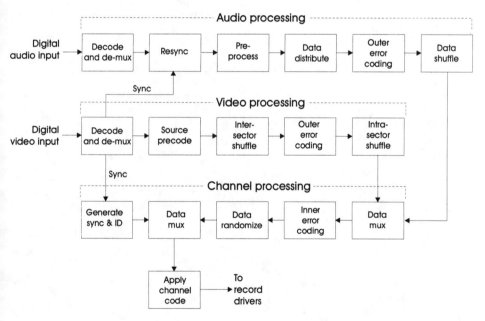

Figure 10.10 D-1 signal processing block diagram.

Table 10.7
D-1 Typical Specifications (NTSC)

Parameter	Value
Video format	4:2:2 Component
Tape width	19 mm (3/4 in)
Cassette	D-1 Standard
Tape speed	286.588 mm/s
Tape usage	54.5 cm^2/s
Channels for video	4
Record heads/channel	1
Total heads on scanner drum	12
Number of audio channels	4
Video track width	0.045 mm
Azimuth recording	No
Drum diameter	75 mm
Drum rotation speed	150 rps
Video channel coding	R-NRZI
Video writing speed	35.3 m/s
Video sampling—luminance	13.5 MHz, 8 bps
Video sampling—chrominance	6.75 MHz, 8 bps
Total video data rate	216 Mbps
Area density	4 Mb/cm^2
Video bandwidth	5.75 MHz
Chrominance bandwidth	2.75 MHz
Video SNR	56 dB
Transient response "K" factor	< 1%
Audio sampling	48 kHz, 20 bps
Audio bandwidth	20 Hz to 20 kHz
Audio dynamic range	105 dB
Audio distortion	< 0.02%
Audio wow and flutter	Unmeasurable

Figure 10.11 *The Sony DVR-28 D-2 composite digital recorder. Photo courtesy of Sony Corporation.*

10.2.1.2 D-2 Composite Recorder

Although a component digital recorder such as D-1 gives excellent performance, it does not integrate well into an otherwise analog video system, which represents most systems today. That is because ADC to a component format must be done every time a signal is recorded, and DAC most be done every time a playback signal goes to an existing analog processor. When dealing with signal sources that are already encoded as NTSC or PAL, overall system performance can be better than analog recording simply by digitizing the composite format and recording that. This actually requires a lower data rate than component encoding, so such recording will also be less expensive. The D-2 composite digital system was designed for that purpose. Figure 10.11 shows a photograph of a current D-2 recorder, the Sony DVR-28.

The D-2 format is defined by the following ANSI/SMPTE standards:

245M	Tape Records
246M	Magnetic Tape
247M	Helical Data and Control Records
248M	Cue Record and Time and Control Code Record
226M	Tape Cassettes (D-1)

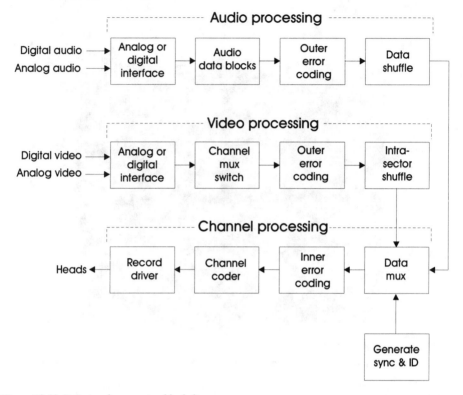

Figure 10.13 *D-2 signal processing block diagram.*

Tape and Cassette—D-2 uses the same cassettes as D-1, but with higher-performance tape inside. This, and the somewhat lower total data rate than D-1, allows longer maximum recording times: S cassette—32 min, M cassette—94 min, L cassette—208 min.

Track Patterns—The track patterns for D-2 are shown in Figure 10.12. Azimuth recording is used in the helical tracks and four audio sectors are provided at the ends of each track. Three longitudinal tracks provide for time code, control, and cue audio recording.

Signal Processing—D-2 signal processing is shown in Figure 10.13. Interfaces are provided for either analog or digitized composite video. The processing is conventional outer and inner error coding plus data blocking and shuffling to achieve the track data format. The channel coding is Miller^2.

Specifications—Table 10.8 shows typical specifications for a D-2 recorder.

Table 10.8
D-2 Typical Specifications

Parameter	Value
Video format	Composite NTSC or PAL
Tape width	19 mm (3/4 in)
Cassette	D-1 Standard (different tape)
Tape speed	131.7 mm/s
Tape usage	25 cm²/s
Channels for video	2
Record heads/channel	2
Total heads on scanner drum	10
Number of audio channels	4
Video track width	0.039 mm
Azimuth secording	Yes
Drum diameter	96.44 mm
Drum rotation speed	89.9 rps
Video channel coding	Miller²
Video writing speed	27.2 m/s
Video sampling	$4f_{SC}$, 8 bps
Total video data rate	115/142 Mbps (NTSC/PAL)
Area density	4.6/5.7 Mb/cm²
Video bandwidth	5.5 MHz
Video SNR	54 dB
Differential gain, phase	2%, 1°
Transient response "K" factor	< 1%
Audio sampling	48 kHz, 20 bps
Audio bandwidth	20 Hz to 20 kHz
Audio dynamic range	105 dB
Audio distortion	< 0.02%
Audio wow and flutter	Unmeasurable

Figure 10.14 *A D-3 recorder and cassettes. Photo courtesy of Panasonic (AJ-D360).*

10.2.1.3 D-3 Composite Recorder

The marketing of composite digital recorders was so successful that a second version was developed using 1/2-in tape, which allows smaller, less expensive equipment. This is D-3, shown in Figure 10.14 and defined in the following ANSI/SMPTE standards:

263M	1/2-in Tape Cassette
264M	D-3 Composite Format—525/60
265M	D-3 Composite Format—625/50

Tape and Cassette—The tape cassettes hold 1/2-in metal-particle tape with the following maximum recording times: S cassette—64 min, M cassette—125 min, L cassette—245 min.

Track Patterns—The D-3 track patterns are shown in Figure 10.15. Audio information for four channels is at the ends of the helical tracks; azimuth recording is used. Three longitudinal tracks record control, time code, and cue audio information.

Signal Processing—Figure 10.16 is a block diagram of D-3 signal processing. This is very similar to D-2 processing except for the use of EFM channel coding and a different blocking strategy.

Specifications—D-3 parameters and specifications are listed in Table 10.9.

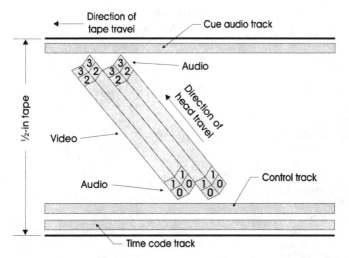

Figure 10.15 *D-3 track patterns.*

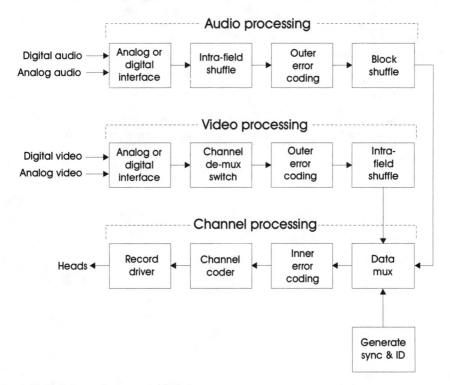

Figure 10.16 *D-3 signal processing block diagram.*

Table 10.9

D-3 Typical Specifications

Parameter	Value
Video format	Composite NTSC or PAL
Tape width	1/2 in (12.65 mm)
Cassette	D-3 Standard
Tape speed	83.88 mm/s
Tape usage	10.6 cm^2/s
Channels for video	2
Record heads/channel	2
Total heads on scanner drum	10
Number of audio channels	4
Video track width	0.02 mm
Azimuth recording	Yes
Drum diameter	76.2 mm
Drum rotation speed	89.9 rps
Video channel coding	EFM
Video writing speed	21.5 m/s
Video sampling	$4f_{SC}$, 8 bps
Total video data rate	115/142 Mbps (NTSC/PAL)
Area density	10.8/13.4 Mb/cm^2
Video bandwidth	5.5 MHz
Video SNR	54 dB
Differential gain, phase	2%, 1°
Transient response "K" factor	< 1%
Audio sampling	48 kHz, 20 bps
Audio bandwidth	20 Hz to 20 kHz
Audio dynamic range	105 dB
Audio distortion	< 0.02%
Audio wow and flutter	Unmeasurable

Figure 10.17 A D-5 recorder. Photo courtesy of Panasonic (AJ-D580).

10.2.1.4 D-5 Component Recorder

It was only natural that there should also be a 1/2-in component format. This is the D-5 standard, defined in the following ANSI/SMPTE Standards:

279M   D-5 Component Format—525/60 and 625/50
263M   Tape Cassette

Figure 10.17 shows a photograph of a D-5 recorder.

Tape and Cassette—The tape cassette and the tape are the same as for D-3; in fact, the compatibility goes so far that D-5 recorders are generally designed to play D-3 tapes.

Track Patterns—D-5 track patterns are shown in Figure 10.18. Four parallel helical tracks are used with azimuth recording to achieve the necessary data rate. Four audio channels are recorded in blocks at the center of the helical tracks. As in the other D-series standards, there are three longitudinal tracks for recording of control, time code, and cue audio information.

Signal Processing—A block diagram of D-5 signal processing is shown in Figure 10.19. The blocks look very similar to D-3 but, of course, the circuits are actually different because of the handling of three components instead of a single digitized composite signal.

Specifications—D-5 parameters and specifications are shown in Table 10.10.

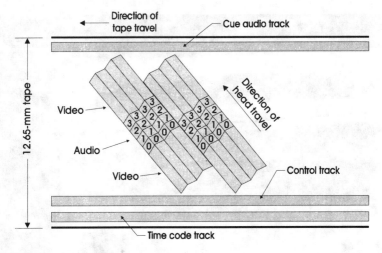

Figure 10.18 *D-5 track patterns.*

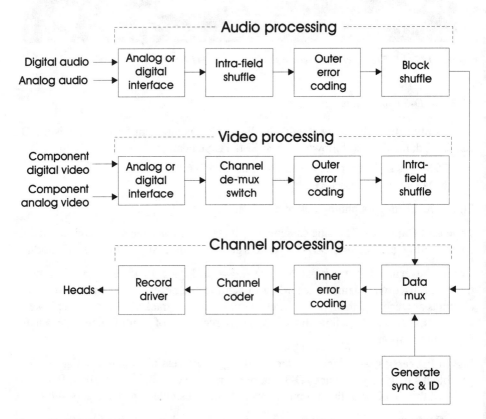

Figure 10.19 *D-5 signal processing.*

Table 10.10
D-5 Typical Specifications

Parameter	Value
Video format	4:2:2 Component
Tape width	12.65 mm (1/2 in)
Cassette	D-3 Standard
Tape speed	167.228 mm/s
Tape usage	21.1 cm^2/s
Channels for video	4
Record heads/channel	2
Total heads on scanner drum	10
Number of audio channels	4
Video track width	0.018 mm
Azimuth recording	Yes
Drum diameter	76.2 mm
Drum rotation speed	89.9 rps
Video channel coding	EFM
Video writing speed	21.5 m/s
Video sampling—luminance	13.5 MHz, 10 bps (NTSC)
	18.0 MHz, 8 bps (PAL)
Video sampling—chrominance	6.75 MHz/9 MHz
Total video data rate	270 Mbps (NTSC/PAL)
Area density	12.8 Mb/cm^2
Video bandwidth	5.5 MHz
Video SNR	54 dB
Transient response "K" factor	< 1%
Audio sampling	48 kHz, 20 bps
Audio bandwidth	20 Hz to 20 kHz
Audio dynamic range	105 dB
Audio distortion	< 0.02%
Audio wow and flutter	Unmeasurable

10.2.1.5 D-6 HDTV Component Recorder

With the imminence of widespread HDTV broadcasting, program creators need high quality recorders for HDTV signals. D-6 is the first standard for that purpose. It records 4:2:2 component signals on 19-mm (3/4-in) tape in various video formats up to 1920 × 1080 active pixels at 30 frames/s. D-6 is defined in the following ANSI/SMPTE standards:

277M   Helical Data, Longitudinal Index, Cue and Control Records
278M   Content of Helical Data and Time and Control Code records
226M   Cassette

Tape and Cassette—The D-6 cassette is the same physical design as the D-1 cassette defined in SMPTE 226M; the tape is metal-particle tape of 1700 oersteds coercivity.

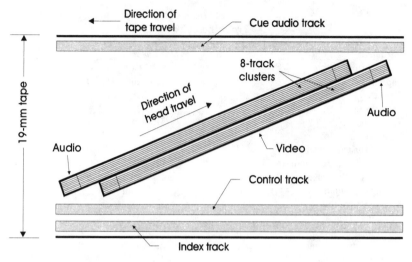

Figure 10.20 *D-6 track patterns.*

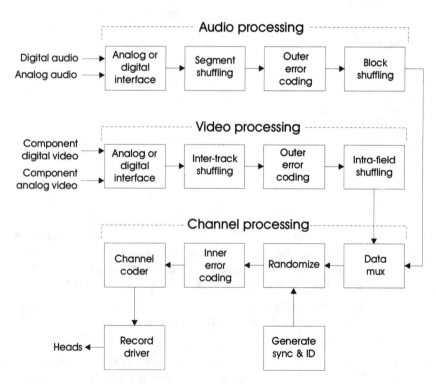

Figure 10.21 *D-6 record signal processing.*

Track Patterns—Track patterns for D-6 are shown in Figure 10.20. Eight helical tracks are recorded in parallel to achieve the data rate of approximately 1.2 Gbps. Audio is recorded in blocks at the ends of the helical tracks; depending on the video standard used, as many as 12 audio channels are available. Three longitudinal tracks provide for index, control, and cue audio recording. The normal format of the index track recording is SMPTE 12M time code.

Signal Processing—Figure 10.21 is a block diagram of the record processing for D-6. This is similar to that of other component recorders except, of course, it expands to the eight parallel head channels when going on and off tape.

Specifications—Parameters and performance specifications for D-6 recorders are shown in Table 10.11. The D-6 system is an expensive solution that provides excellent performance. However, many users would accept the performance trade-off of compression to reduce costs. Such systems are being developed. An alternative approach is to produce an adaptor to record compressed HDTV on an SDTV composite digital recorder, such as D-3. This also is being developed.

Table 10.11
D-6 Typical Specifications

Parameter	Value
Video format	4:2:2 Component HDTV
Tape width	19 mm (3/4 in)
Cassette	D-1 Standard
Tape speed	497.4 mm/s
Tape usage	94.5 cm^2/s
Channels for video	8
Record heads/channel	2
Total heads on scanner drum	34
Number of audio channels	10–12
Video track width	0.021 mm
Azimuth recording	Yes
Drum diameter	96.52 mm
Drum rotation speed	150 rps
Video channel coding	8,12
Video writing speed	45.5 m/s
Video sampling—luminance	74.25 MHz (typical)
Video sampling—chrominance	38.12 MHz (typical)
Total video data rate	1.2 Gbps (maximum)
Area density	12.7 Mb/cm^2
Video bandwidth	—
Video SNR	—
Transient response "K" factor	—
Audio sampling	48 kHz, 20 bps
Audio bandwidth	20 Hz to 20 kHz
Audio dynamic range	105 dB
Audio distortion	< 0.02%
Audio wow and flutter	Unmeasurable

Figure 10.22 *Typical DV hand-held camcorder. Photo courtesy of Panasonic (AG-EZ1).*

10.2.2 DV

The digital standards described so far do not use video compression beyond that implied by 4:2:2 sampling. For production-postproduction use, it is desirable to have little or no compression simply to avoid the signal degradation that may occur with it, and that may accumulate with repeated compression-decompression cycles in multiple generations. However, many markets are not so demanding and substantial cost reduction can be achieved by using compression. The extreme case of this is, of course, the home market for camcorders, where cost alone may determine whether a market exists or not. Here, digital video recording with compression can offer much more picture and sound quality than analog recorders, and the price point of such products is approaching the level that will allow mass marketing. The remarkable feature of these products is that they are size-competitive with analog camcorders; that is a result of IC technology, of course, which is eminently applicable to digital circuits.

The standard for home digital recording is called *DV*, described in this section. DV uses 1/4-in tape in a cassette that provides up to 4 hrs of recording. Because of the small size and low cost of DV components, DV has also spawned several professional standards that trade playing time against the amount of compression; two of these are described in later sections.

10.2.2.1 DV Tape Decks

A photograph of a typical DV hand-held camcorder product is shown in Figure 10.22.

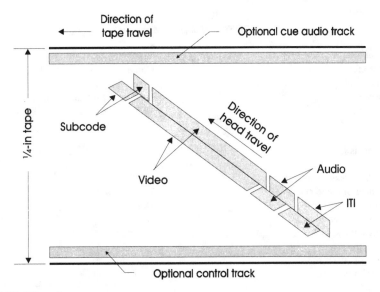

Figure 10.23 *DV track patterns.*

10.2.2.2 DV Track Patterns

DV track patterns are shown in Figure 10.23. Helical tracks are recorded using azimuth recording. Blocking of the helical tracks provides for audio, subcode, and insert-and-track-information (ITI) code. The subcode block contains time code and other information that can identify tracks in high-speed shuttle modes. Two longitudinal tracks are available but they are both optional—the system can operate without them.

10.2.2.3 DV Signal Processing

The important difference between this system and the others described so far is the use of video compression. Component video is sampled at 13.5 MHz 4:1:1 for 525 lines or 13.5 MHz 4:2:0 for 625 lines. This gives a data rate of about 120 Mbps, which is compressed 10:1 using DCT for the signal to be recorded.

10.2.2.4 DV Specifications

Parameters and specifications for DV are given in Table 10.12.

Table 10.12
DV Typical Specifications

Parameter	Value
Video format	Component, compressed 10:1
Tape width	1/4 in (6.35 mm)
Cassette	DV Standard, small or large
Tape speed	19.05 mm/s
Tape usage	1.23 cm²/s
Playing time	2 hr, 4 hr
Channels for video	1
Record heads/channel	2
Total heads on scanner drum	4
Number of audio channels	2 (in helical tracks)
Audio sampling rate	48 kHz
Audio bits/sample	16
Helical track width	0.01 mm
Azimuth recording	Yes
Drum diameter	21.7 mm
Drum rotation speed	150 rps
Helical writing speed	10.2 m/s
Video sample rate—Y	13.5 MHz
Video sample rate—C_R, C_B	6.75 MHz
Video decimation	4:1:1 (525), 4:2:0 (625)
Video compression	DCT plus RLC
Total video data rate after compression	12 Mbps
Area density	9.8 Mb/cm²

10.2.3 DVCPRO

As with 1/2-in tape systems, the 1/4-in consumer design has been adapted to higher-performing professional recorders. Two are described here—DVCPRO by Panasonic (this section) and DVCAM by Sony (see next section).

The 10:1 compression used by DV is too much for professional use; in DVCPRO, the compression is reduced to 5:1, giving better picture quality at a data rate of 25 Mbps. Intra-frame compression is used to avoid editing restrictions. Recording of this is achieved by increasing the helical track width by raising the tape speed, and by using better tape, but the scanner diameter and rotational speed remain the same as in DV. A second version, DVCPRO50, reduces the compression still further, to 3.3:1 to achieve recording of full 4:2:2 sampling. The data rate for DVCPRO50 is 50 Mbps.

DVCPRO products offer interfaces for analog component, analog Y/C, analog composite, or serial digital (SDI) video. Audio interfaces also are provided for analog or digital.

10.2.3.1 DVCPRO Tape Decks

Figure 10.24 shows photographs of several DVCPRO products. As with the other profes-

Figure 10.24 DVCPRO products: (a) Studio VCR, (b) Camcorder, (c) Field Editor, and (d) Dockable VCR. Photos courtesy of Panasonic.

sional digital recording systems, a complete product line for production and postproduction is available, including camcorders, field editors, editing VCRs, edit controllers, and so on. A special recorder is also available that is capable of playing or recording at 4× normal speed. This is useful for fast transfers between DVCPRO tapes and hard disk recorders.

10.2.3.2 DVCPRO Track Patterns

The DVCPRO track patterns are the same as for DV (Figure 10.23) except, of course, the helical tracks are wider. Because of this similarity, DVCPRO units can play DV tapes, which is useful when shots taken with the small DV camcorder need to be used in the DVCPRO system. It can also play Sony DVCAM tapes.

10.2.3.3 DVCPRO Signal Processing

Signal processing for DVCPRO is similar to the D-series recorders except for the addition of compression for both video and audio.

10.2.3.4 DVCPRO Specifications

DVCPRO and DVCPRO50 specifications are shown in Table 10.13.

Table 10.13
DVCPRO and DVCPRO50 Typical Specifications

Parameter	DVCPRO	DVCPRO50
Video format	Component, compressed 5:1	Component, compressed 3.3:1
Tape width	1/4 in (6.35 mm)	1/4 in (6.35 mm)
Cassette	DV Standard, small/large	Same
Tape speed	33.82 mm/s	67.64 mm/s
Tape usage	2.15 cm^2/s	4.3 cm^2/s
Maximum playing time	63/123 min	45/90 min
Channels for video	1	1
Record heads/channel	2	4
Total heads on scanner drum	4	8
Number of audio channels	2 (in helical tracks)	4 (in helical tracks)
Audio sampling rate	48 kHz	48 kHz
Audio bits/sample	16	16
Audio dynamic range	80 dB	80 dB
Helical track width	0.018 mm	0.018 mm
Azimuth recording	Yes	Yes
Drum diameter	21.7 mm	21.7 mm
Drum rotation speed	150 rps	150 rps
Helical writing speed	10.2 m/s	10.2 m/s
Video sample rate—Y	13.5 MHz	13.5 MHz
Video sample rate—C_R, C_B	3.375 MHz	6.75 MHz
Video decimation	4:1:1 (525), 4:2:0 (625)	4:2:2
Video compression	DCT plus RLC	DCT plus RLC
Video bandwidth—Y	5.5 MHz	5.5 MHz
Video bandwidth—C_R, C_B	1 MHz	2.5 MHz
Video SNR (digital-analog)	60 dB	60 dB
Video K-factor	2%	1%
Compressed video data rate	25 Mbps	50 Mbps
Area density	11.6 Mb/cm^2	11.6 Mb/cm^2

10.2.4 DVCAM

The Sony adaptation of DV to professional use is called *DVCAM*™. A complete line of products for production and postproduction is available in this format. Some of the products are illustrated in Figure 10.25. DVCAM products provide analog video interfaces in composite and S-video formats, digital video SDI interfaces, and analog and digital audio interfaces.

10.2.4.1 DVCAM Track Patterns

The DVCAM track patterns are shown in Figure 10.26. Note that no longitudinal tracks are used. This is necessary because the system uses metal evaporated tape, which is not

Figure 10.25 *Photographs of Sony DVCAM products: (a) Studio VCR, (b) Dockable VCR, (c) Field Editing VCR, and (d) Camcorder. Photos courtesy of Sony Corporation.*

suitable for longitudinal tracks. However, the system operates successfully without the longitudinal tracks. The DVCAM tracks are slightly narrower than DVCPRO, which limits the compatibility between those two systems. DVCAM can play DV tapes, however.

10.2.4.2 DVCAM Signal Processing

DVCAM operates with 4:2:0 sampling and intra-frame compression using DCT. The recorded data rate is 25 Mbps.

10.2.4.3 DVCAM Specifications

DVCAM specifications are shown in Table 10.14.

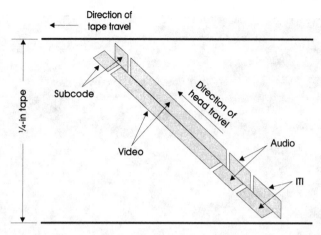

Figure 10.26 *DVCAM track patterns.*

Table 10.14
DVCAM Typical Specifications

Parameter	Value: DVCAM
Video format	4:2:0 Component, compressed 5:1
Tape width	1/4 in (6.35 mm)
Cassette	DVCAM, mini/standard
Tape speed	33.8 mm/s
Tape usage	2.15 cm^2/s
Maximum playing time	40/184 min
Channels for video	1
Record heads/channel	2
Total heads on scanner drum	4
Number of audio channels	2 or 4
Audio sampling rate	2ch–48 kHz, 4ch–32 kHz
Audio bits/sample	2ch–16, 4ch–12
Audio bandwidth	2ch–20 kHz, 4ch–14.5 kHz
Audio dynamic range	80 dB
Helical track width	0.015 mm
Azimuth recording	Yes
Drum diameter	21.7 mm
Drum rotation speed	150 rps
Helical writing speed	10.2 m/s
Video sample rate—Y	13.5 MHz
Video sample rate—C_R, C_B	6.75 MHz
Video decimation	4:2:0
Video compression	DCT
Video bandwidth—Y	5.5 MHz
Video bandwidth—C_R, C_B	2 MHz
Video SNR (digital-analog)	55 dB
Video K-factor	2%
Compressed video data rate	25 Mbps
Area density	11.6 Mb/cm^2

Figure 10.27 *Photograph of a Digital Betacam studio recorder, the Sony DVW-A500. Photo courtesy of Sony Corporation.*

10.2.5 Digital Betacam

Sony's highest-quality 1/2-in digital VTR format is *Digital Betacam™*. Recording is 4:2:2 component with approximately 1.6:1 compression via DCT. Compression is intra-frame, so editing is not limited. A complete line of camcorders, field and studio recorders, and editing recorders is available. Figure 10.27 shows a photograph of a typical Digital Betacam studio recorder.

10.2.5.1 Digital Betacam Track Patterns

Digital Betacam track patterns are shown in Figure 10.28. Helical tracks are recorded by two heads, with blocks for four audio channels at the center of the tracks. Three longitudinal tracks provide for time code, control, and cue audio recording.

10.2.5.2 Digital Betacam Specifications

Specifications for Digital Betacam are in Table 10.15.

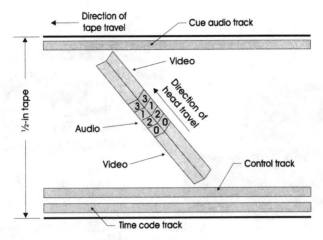

Figure 10.28 *Digital Betacam track patterns.*

Table 10.15
Digital Betacam Typical Specifications

Parameter	Value
Video format	4:2:2 Component, compressed 1.6:1
Tape width	1/2 in (12.65 mm)
Cassette	Digital Betacam, mini/standard
Tape speed	96.7 mm/s
Tape usage	12.2 cm^2/s
Maximum playing time	40/124 min
Channels for video	1
Record heads/channel	2
Total heads on scanner drum	4
Number of audio channels	4
Audio sampling rate	48 kHz
Audio bits/sample	20
Audio bandwidth	20 kHz
Audio dynamic range	100 dB
Helical track width	0.026 mm
Azimuth recording	Yes
Drum diameter	74.4 mm
Drum rotation speed	180 rps
Helical writing speed	16.6 m/s
Video sample rate—Y	13.5 MHz
Video sample rate—C_R, C_B	6.75 MHz
Video decimation	4:2:2
Video compression	DCT
Video bandwidth—Y	5.75 MHz
Video bandwidth—C_R, C_B	2.75 MHz
Video SNR (digital-analog)	62 dB
Video K-factor	1%
Compressed video data rate	125 Mbps
Area density	10.2 Mb/cm^2

DNW-90
Camcorder

DNW-A75
Digital Video Cassette
Recorder

DNV-5
Dockable VTR

DNW-A225
Digital Portable
Editor

Figure 10.29 Sony Betacam SX products. Photos courtesy of Sony Corporation.

10.2.6 Betacam SX

Sony developed a second 1/2-in system to address lower cost professional markets. The *Betacam SX™* system also uses 4:2:2 sampling, but it has a greater compression ratio of 10:1 to reduce tape usage and system cost, particularly for news-gathering systems. At that compression ratio, it is necessary to use inter-frame compression, which places some restrictions on editing, but they are not considered severe for many users. A unique feature of this system is that tapes can be played at up to 4× normal speed without losing any data, which is valuable when transferring video into a nonlinear editing system or when transmitting video via satellite from a new site to the studio. Another valuable feature is the ability to play tapes recorded with the previous analog Betacam products.

Figure 10.29 shows some of the Betacam SX products.

10.2.6.1 Betacam SX Track Patterns

Track patterns for Betacam SX are shown in Figure 10.30. Helical tracks have four audio channels (actually, each are recorded twice in every track) and two sectors for system data. Three longitudinal tracks provide for time code, control, and auxiliary information.

10.2.6.2 Betacam SX Specifications

Specifications for Betacam SX are shown in Table 10.16.

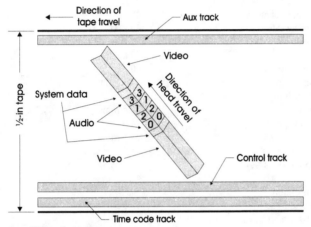

Figure 10.30 Betacam SX track patterns.

Table 10.16
Betacam SX Typical Specifications

Parameter	Value
Video format	4:2:2 Component, compressed 10:1
Tape width	1/2 in (12.65 mm)
Cassette	Betacam SX, mini/standard
Tape speed	59.575 mm/s
Tape usage	7.53 cm^2/s
Maximum playing time	60/184 min
Channels for video	1
Record heads/channel	2
Total heads on scanner drum	8
Number of audio channels	4
Audio sampling rate	48 kHz
Audio bits/sample	16
Audio bandwidth	20 kHz
Audio dynamic range	90 dB
Helical track width	0.032 mm
Azimuth recording	Yes
Drum diameter	74.4 mm
Drum rotation speed	74.925 rps
Helical writing speed	17.5 m/s
Video sample rate—Y	13.5 MHz
Video sample rate—C_R, C_B	6.75 MHz
Video decimation	4:2:2
Video compression	DCT
Video bandwidth—Y	5.75 MHz
Video bandwidth—C_R, C_B	2.75 MHz
Video K-factor	1%
Compressed video data rate	18 Mbps
Area density	2.4 Mb/cm^2

10.3 HDTV RECORDING

The only HDTV recorder described here is the D-6, which records uncompressed component HDTV at data rates up to 1.5 GBps. This is an expensive process and there is a need
for lower cost approaches; the answer, of course, is to use some compression. Two products are on the market and more are coming. The two already available are a conversion of
D-5 by Panasonic, called HD D5, and a conversion of Digital Betacam by Sony, called
HDCAM. Another product announced but not available at this writing is DVCPRO 100
by Panasonic and obviously based on the other DVCPRO products. Table 10.17 gives
some of the specifications for the first two products mentioned.

Table 10.17

Typical Specifications of Compressed-HDTV Recorders

Parameter	Panasonic HD D5	Sony HDCAM
Video format	HDTV 4:2:2, compressed 4:1	HDTV 3:1:1, compressed 4.4:1
Video bits/sample	10	8
Tape width	1/2 in (12.65 mm)	1/2 in (12.65 mm)
Cassette	D-3 S/M/L	Digital Betacam S/M/L
Tape speed	167.228 mm/s	96.7 mm/s
Tape usage	21.1 cm^2/s	12.2 cm^2/s
Maximum playing time	23/63/124 min	40/60/124 min
Number of audio channels	4 (in helical tracks)	4 (in helical tracks)
Audio sampling rate	48 kHz	48 kHz
Audio bits/sample	20	20
Helical track width	0.018 mm	0.022 mm
Azimuth recording	Yes	Yes
HD video sample rate—Y	74.25 MHz	55.7 MHz
HD video sample rate—C_R, C_B	37.125 MHz	18.56 MHz
Video compression	Intrafield	Interfield
Video bandwidth—Y	5.5 MHz	5.5 MHz
Video bandwidth—C_R, C_B	1 MHz	2.5 MHz
Compressed video data rate	270 Mbps	140 Mbps
Area density	12.8 Mb/cm^2	11.5 Mb/cm^2

10.4 CONCLUSION

This chapter has given a sampling of the recording products on the market. It is by no means complete and there are other formats and other manufacturers not covered here. Especially in the consumer market, there are many manufacturers building the same formats and there are many variations of product design as the manufacturers vie for market share.

REFERENCE

[1] Luther, A. C., *Principles of Digital Audio and Video*, Artech House, Norwood, MA, 1997.

11

Disk Recorder Products

Video recorders store their information on two-dimensional media such as tapes or disks. For a given recording technology, the area density factor (see Section 2.1.2) defines the technology's efficiency in using the medium, and, for a given area density, the more area available, the more recording time is possible. With early video recorders, the area density capability was low and the only way to achieve reasonable recording times was to use the large surface area available from a tape. Disk media, whose recording area is limited to that of the disk surfaces, could not reach playing times of more than a few seconds of video.

However, that has changed with the development of digital optical recording methods. Area density performance now allows 1 hour or more of high-quality video and audio to be stored on a single surface of a 12-cm disk. That means the inherent advantage of disks over tapes can be exploited for rapid random access to any position in a recording, which is a fundamental need in editing systems.

A parallel development has occurred in magnetic mass storage devices for computers. Storage capacities of hard disk drives are now at a level that allows the recording of high-quality digital video up to 1 hour or more per drive. This, coupled with the capability of a computer, is the basis for the rapid growth in the use of nonlinear editing systems in all areas of video program creation. This chapter discusses some of the disk equipment for recording video and audio.

11.1 THE COMPACT DISC

Although the analog laser videodisc was the first mass-produced optical recording application (see Section 11.4), the Compact Disc (CD) for distribution of sound recordings to the home was the first all-digital optical recording product. Subsequent application of CD technology to personal computers, as well as the product improvements that have resulted from the impetus of both home and computer markets, has led to the use of this

241

technology for many digital video applications. CD technology is described in Sections 2.3.4.1, 3.3.3, 4.3.1, and 6.3; here, some of the equipment available with CD technology will be described.

The ubiquitous CD audio player has been manufactured in almost all imaginable configurations, from stereo component cabinets to hand-held portable units. Prices have fallen to rock bottom levels so that almost anyone can afford a CD-audio player. The technology and manufacturing developments that make possible this packaging diversity and low cost are directly applicable to the computer and video uses of this technology.

11.1.1 CD-ROM

The CD-ROM resulted from taking the basic CD-audio mechanism and interfacing it to a computer as a means to replay digital recordings distributed by external sources. This has proved to be extremely valuable for the distribution of computer software, data libraries, and sometimes audio and video for computers. From the viewpoint of computers, however, there were serious limitations in the early CD-ROM players based on adapting the audio players.

- The data rate of 1.2 Mbps limited the computer performance.
- The access time was as much as 1 second.
- Recording was only possible with expensive equipment that was not practical to place in an ordinary computer.

Nevertheless, as the market for CD-ROM drives grew to the point that they became standard in nearly all PCs, the industry addressed these limitations and significant advances have been made.

11.1.1.1 CD-ROM Tracks

The CD-ROM is recorded with a single physical track that spirals from the inner recording radius to the outer radius. Adjacent tracks in the spiral are spaced 1.6 μm apart and the tracks have a width of 0.6 μm. Digital data are recorded with EFM encoding in frames of 588 channel bits; a frame contains overhead bits to support clock extraction, synchronization, preliminary error correction, and track identification. The net data content of a frame is 24 input bytes.

For logical purposes, CD-ROM data are divided into 2,352-byte blocks or sectors; these are the minimum addressable units for random access. This is specified by the so-called *Yellow Book* standard, which provides two data modes for sectors. In general, an entire CD-ROM operates in one or the other mode, although there are also several subordinate standards that can support *mixed-mode* discs or even interleaved modes.

Figure 11.1 shows the usage of the 2,352-byte sector contents for the two modes. The difference is in the amount of error correction: Mode 1, which is used with computer data, provides a corrected error rate of 10^{-12}, whereas Mode 2, which is used for digital audio or video, delivers an error rate of only 10^{-9}. This recognizes the fact that occasional errors in

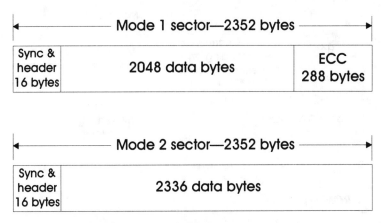

Figure 11.1 *CD-ROM sector contents for Mode 1 and Mode 2.*

audio or video can be effectively concealed, but a single error in critical computer data may be catastrophic. So, for computer data, a Mode 1 sector holds only 2,048 bytes of information because 288 bytes are assigned to error correction codes (ECC). This is called *layered* error correction because it is on top of the basic ECC contained in the channel coding. A Mode 2 sector can hold 2,336 bytes of audio or video data because there is no layered ECC. The original *Red Book* CD-audio standard uses all 2,352 bytes of the sector for audio, relying only on the error protection contained in the channel coding for data integrity.

Although the data modes of the *Yellow Book* standard provide for recording and replaying data, they do not include any description of how a computer will find a particular segment of data on a CD-ROM. This feature is provided by a *file system* standard, which defines how segments of data (files) on the CD are given names and how directories are set up so the locations of all files can be read from the disc. The CD-ROM file system is defined by the *ISO 9660* standard, which is used by nearly all CD-ROM producers.

11.1.1.2 CD-ROM Signal Processing

The signal processing of a CD-ROM drive is diagrammed in Figure 11.2. Signals from the optical head are amplified and delivered to the servo system for tracking and focus control. The channel coding of the data signal is decoded and error correction applied. In Mode 2 (CD-ROM audio or video), this becomes the data output, which is usually buffered in the CD-ROM drive for retrieval at the convenience of the host system. When Mode 1 (general data) is used, the layered error processing is performed before passing the data on to the I/O buffer.

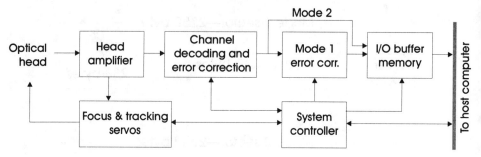

Figure 11.2 *Signal processing of a CD-ROM player.*

11.1.1.3 CD-ROM Data Rate

One of the first CD-ROM improvements beyond CD-audio was to speed up the data rate, which required faster rotation of the disc. Since the CD operates in CLV mode (see Section 3.4.1.2) for the audio CD data rate, the rotational speed ranges from 200 rpm at the outer radius to 530 rpm at the inner recording radius. This is known as *single-speed* or 1× operation.

Improved drives operate at a multiple of single-speed, such as 2×, 4×, 8×, and so forth. Maximum speeds up to 32× are now available. This requires a speed range from 6,400 to 16,960 rpm; however, the high end of that range is not achievable, so faster drives are known as *variable-speed*, with their highest speed multiple only being achieved at the outer radius. The rotation speed is not increased by as large a factor at the inner radii, so the speed multiple drops. For video applications, where a constant or nearly-constant data rate is required, the inner-radius data rate will be the governing factor. For example, if the so-called 32× drive mentioned above maintains its maximum rotation at 6,400 rpm on inner radii, it will only be a 12× drive for video, giving a data rate of 14.4 Mbps.

Since there is a fixed amount of recording space on a CD-ROM, and the track pattern is also fixed, speeding up disc rotation to obtain higher data rates will reduce playing time. A 1× disc has a maximum playing time of 74 min; the hypothetical 12× disc described above will play only a little more than 6 min. Thus, longer playing times at high data rates require an increase in disc capacity, which was the major objective of the DVD program (see Section 11.2).

11.1.1.4 CD-ROM Access Time

The access time of early 1× CD-ROMs was in the range of 1 second. This was determined by the speed of head movement across tracks, the tracking servo lock-up that must occur when changing tracks, and the slow disc rotational rate. As rotational speeds were increased, there was some improvement in access time, but it was also necessary to redesign head positioners and servo systems to reduce their contribution to access time. Even so, access times less than about 100 ms are rare, even in the fastest drives. This compares to hard disk access times, which range below 10 ms.

Most PCs do not keep high-speed CD drives rotating at full speed all the time. When the CD is accessed the first time, the disc will spin up to speed, which takes a second or two. However, data does get read while speed-up is going on, so the first access takes somewhat longer than later ones, after the disc is up to speed. If the drive is not accessed for a period of time, the drive controller allows the disc to slow down again. The reason for this form of speed control is to reduce acoustic noise, save power, and reduce bearing wear.

11.1.1.5 Recordable CD-ROM (CD-R)

Mastering (recording) of audio CDs requires chemical processing of photoresist, which is exposed by a laser beam and processed to create a master disc. This is then subjected to electroplating to create stamper discs that are used to replicate CDs by pressing on plastic substrates. It is a complex process that can only be done in a dedicated manufacturing plant. However, the utility of a CD-ROM drive in a PC would be greatly enhanced if CDs could be recorded as well as read.

This goal of recording CDs or CD-ROMs one at a time in a PC has been achieved with a process that simply involves writing a CD with a laser beam in a drive that is similar to a CD player. This is known as *CD-R*. Its secret is an organic dye material for the recording layer that locally changes its optical properties when exposed to a laser beam of moderate power. Recordings can be made that are playable on any CD player device. However, CD-R is a write-once recording method, meaning that a recording cannot be overwritten or modified after it is once written (see Section 3.3.3.2).

Because of the need for a given amount of laser power per unit time to accomplish recording, CD recording cannot tolerate as much speed-up as playing. Most CD recorders operate at 2× speed or, at most, 4×. There are also limitations to be considered with regard to how rapidly the PC can provide the data to be recorded; the CD recorder must receive data smoothly without interruptions. Although the CD recorder itself may contain some data buffering to accommodate small input interruptions, if the recorded data ever is actually interrupted, the recording is ruined. Thus, the PC data delivery must operate conservatively to be sure an interruption through the buffering never occurs.

11.1.1.6 Video CD

The data rate of a 1× player is 1.2 Mbps. That is sufficient for MPEG-1 video, but nothing more. Although the fastest PCs can decompress MPEG-1 in software, it helps to have hardware support for the task, especially if the video is to be displayed full-screen. For this reason, many current PC video display adaptors include MPEG-1 decompression features.

A standard for *Video CD* was developed using MPEG-1. It was expected that dedicated (non-PC) players would be built for the home market and this would subsume some of the market now enjoyed by prerecorded videotapes. However, although CDs are more convenient and reliable than videotape, in the Video CD form they do not offer any performance advantage over videotape and have not seen much acceptance. The industry is now pursuing the same objective with the DVD system, which offers significant improvements in both audio and video performance.

As already explained, higher quality compressed video, such as MPEG-2, can be played by a faster CD-ROM drive with a trade-off of playing time. However, this trade-off is too severe for most applications, and the capability is rarely used. Applications that need to include significant amounts of MPEG-2 video are looking to the DVD standard described below.

11.1.1.7 Rewritable CD (CD-RW)

A CD is made rewritable (rerecordable) by using a recording layer where records can be erased and rerecorded by application of suitable laser power levels. One type of material is a crystalline alloy layer that is initially highly reflective, but at a specific laser power level, the crystalline layer converts to an amorphous state (a phase change) that has low reflectivity (see Section 3.3.3.2). This can be detected by the normal CD player optics. The phase change process is reversible—the layer can be returned to the crystalline reflective state by applying a slightly lower laser power than that used for recording. Thus, such a recorder has three levels of laser power: play (lowest), erase (higher), recording (highest).

CD-RW discs have a grooved structure similar to CD-R blanks. Owing to the critical nature of the rewritable material, they also contain prerecorded data tracks that provide the proper recording and erasing parameters for each particular disc. CD-RW drives must read this information and set themselves up for optimal recording on each disc.

Although CD-RW discs can nominally be played by any CD drive, the recording contrast (the difference in reflectance between reflective and nonreflective areas) is lower, and drives must contain an AGC feature in the playback channel to accommodate normal CDs or CD-RW discs. This is included in new drives, but older drives now in service may not be able to read all CD-RW discs.

11.1.1.8 CD-ROM Drives

Packaging of CD-ROM drives is constrained by the systems in which they are used. In a desktop PC, CD-ROM drives are inserted into drive bays on the front of the cabinet, that are of dimensions dating from the days of 5¼-in (13.33-cm) floppy disks. These bays are 14.76 cm wide and generally support varying drive heights. However, the so-called *half-height* drive is most common, having a height of about 2.9 cm.

The front of a CD-ROM drive has a door for insertion of the CD-ROM disc plus an eject button for getting a disc back out of the drive when it is no longer needed in the system. Optionally, a drive may have an *activity light* that indicates when the system is accessing the disc, an earphone jack for listening to CD-audio discs, and a volume control for the earphone.

There are two methods of loading CD-ROM discs into drives—the drawer or the caddy. The drawer approach is the same as used in most CD-audio drives, where a drawer opens either automatically or manually and the disc is placed in the drawer, which then is drawn back into the drive. This is the preferred approach today.

Early CD-ROM drives used the *caddy* approach, where the disc was placed into a carrier (caddy) outside the drive, and the caddy was then inserted into a slot at the front of the drive. This was intended to provide better protection to the disc (which it does), but the additional handling required, the cost of caddies, and the inherent durability of the medium makes it not worth the trouble.

A second packaging for a CD-ROM drive was needed for use with portable PCs. Here, considerations of size, weight, and power consumption outweigh the cost factors. A number of smaller packages have been developed for this market, although the ultimate limiting factor is still the size of the disc itself. These drives all use the drawer loading approach; sometimes the drawer is the entire drive, which comes out of the machine, or the machine opens up to the drive for insertion of a disc.

11.2 DIGITAL VERSATILE DISC

It shouldn't be surprising that after producing hundreds of millions of CD-audio and CD-ROM drives and billions of discs over the last 15 years, that things have been learned to improve the performance of these systems, if standards were to be changed. The *Digital Versatile Disc* (DVD) is the industry's choice of a new standard for optical discs that vastly increases the data storage capacity, while adding new features that should help keep the standard alive for the next 15 or 20 years. One important impetus for the DVD standard was to achieve a digital video capability that would improve audio and video performance well beyond that of prerecorded videotapes, so as to have the potential to take over the prerecorded video market. Those performance goals have been achieved in the DVD standard, which is seeing initial market deployment at the time of this writing.

The performance advances of DVD have been achieved without any major techno-logical breakthrough, except maybe the multilayer technology that allows up to four re-cording layers to be included on a single plastic substrate. The basic improvement in one-layer storage capacity has been obtained simply by tightening up on the track patterns to reach higher density. Of course, this requires incremental improvements in optics, lasers, photodetectors, and servo systems; but the resulting hardware bears a marked similarity to the original CD-audio hardware.

11.2.1 DVD Tracks

Track pitch of DVD has been reduced to 0.74 µm, a factor of 2.2 smaller than CD-ROM. Minimum pit length is reduced to 0.4 µm, a factor of 2.1 less than CD-ROM; this requires a reduction of laser wavelength to 650 nm (from 780 nm for the CD) and an increase in the objective lens numerical aperture to 0.6 (from 0.45 for CD players). These factors give a 4.6× increase in data density compared with CD-ROM, but additional increases are achieved by a slight increase in the recording area, improved channel coding, and reduced error

Table 11.1

DVD Data Capacity and Playing Times

Sides/layers	Capacity (GB)	Video playing time (min)*
Single/single (SS/SL)	4.38	133
Single/double (SS/DL)	7.95	241
Double/single (DS/SL)	8.75	266
Double/double (DS/DL)	15.90	483

* Based on 3.5 Mbps MPEG-2 SDTV video plus 1.15 Mbps for three channels of Dolby Digital Audio.

protection overhead. The result is a 7× increase in storage capacity of a single layer—from 680 MB to 4.38 GB.

Channel coding for DVD is eight-to-sixteen modulation (sometimes called *EFMPlus*), which is slightly more efficient than CD-ROM, which uses actual EFM with three merging bits added between words, resulting in a net 8/17 modulation.

DVD offers additional storage capacity through its *multilayer* capability. Both sides of a disc can be used and each side can have two recording layers, selectable by refocusing the reading laser beam. The playing times of the various combinations of layers and sides are shown in Table 11.1. The table shows that the capacity of the second layer is slightly less than the first layer; this is because the density has to be reduced slightly on the second layer to account for the loss of reading through the first layer. A full DS/DL disc can store more than 8 hours of SDTV video. Note that there is nothing magic about the data rates used in Table 11.1; both audio and video data rates or number of channels can vary as long as the total rate remains below the DVD maximum of 10.08 Mbps. Of course, playing times will shorten or lengthen according to the data rates used.

The first layer on a DVD is recorded from inner to outer radius. When a second layer is used, recording may be done in either direction. Recording the second layer from outer to inner radius allows a very rapid continuation of playback from the first layer; since the head does not have to be moved, only the laser is refocused to the second layer.

11.2.2 DVD Sectors

DVD does not have the equivalent of the CD-ROM Modes 1 and 2. All sectors are designed for general data, whether the content is computer data, audio, or video. The sector structure is shown in Figure 11.3. Although user data is partitioned into 2,048-byte logical sectors, the error protection system of DVD operates on a block comprising 16 sectors with 5,088 bytes of overhead added. Thus, data must be read from the disc in 37,856-byte blocks for the purpose of error detection and correction. When using random access, buffering in the drive allows individual logical sectors to be extracted from the error protection blocks, so that access to any sector is possible. The error protection uses Reed-Solomon product coding (RS-PC) to achieve a corrected error rate of 10^{-15} or better. This is more than a 1000× improvement over CD-ROM error protection.

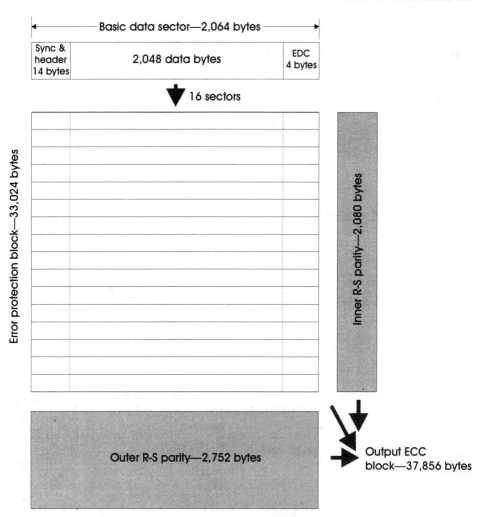

Figure 11.3 DVD sector structure.

11.2.3 DVD Signal Processing

Basic signal processing of a DVD drive is shown in Figure 11.4. Since DVD drives are required to also read CD-audio discs, there are two processing circuits in every drive. The digital data from the DVD data processor may be processed further depending on the application of the drive. For example, a DVD-Video drive will include circuits for demultiplexing and decoding the compressed audio and video streams contained in the data (see Section 11.2.5).

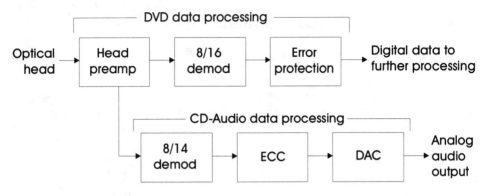

Figure 11.4 *Signal processing of a DVD player.*

11.2.4 DVD-ROM

The CD was initially designed as an audio medium and the CD-ROM and other formats came later; but in recognition of the importance and generality of the computer application, DVD was initially defined as a ROM medium. From this base, other applications are defined.

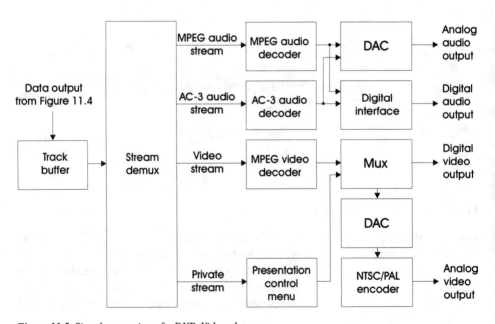

Figure 11.5 *Signal processing of a DVD-Video player.*

11.2.5 DVD-Video

One of the DVD applications is DVD-Video, intended for distribution of prerecorded video and audio programs to homes and other users. The DVD video standard supports MPEG-1 and MPEG-2 video and Dolby Digital or MPEG audio.

Figure 11.5 shows the additional processing required in a DVD-Video player beyond that of Figure 11.4. The error-corrected data from the disc flow at a constant data rate of 11.08 Mbps into a track buffer. The output of the buffer contains several packetized elementary streams, which may include audio, video, subpicture data, presentation control information (PCI), or data search information (DSI). Decoder modules in the player process the various streams and present them as analog or digital outputs. The PCI and DSI streams are used by the player to display menus to the user and to locate files on the disc.

Many other features are provided in the DVD-Video standards, including auxiliary files on the disc to display still images, navigation menus, titles, and so on.

DVD-Video players generally have analog outputs to connect to analog TV receivers. As digital TVs become available, DVD players may also have digital outputs. Figure 11.5 shows both.

11.2.6 DVD-RAM

The same recording technologies used in CD-R and CD-RW can be used with DVD to produce recordable and rewritable versions. These have been considered in the standards and will appear much earlier in the evolution of DVD than they did with CD simply because the underlying technologies for recording now exist whereas they didn't when the CD first came out.

The erasable-recordable format of DVD-ROM has been named DVD-RAM. Although sample drives are available at this writing, standards have not been settled, so serious marketing has not begun. It is already clear, however, that early DVD-ROM drives cannot read the discs produced by these drives, and a small change will be needed in DVD-ROM drives to accommodate DVD-RAM discs. Also, DVD-RAM discs will have less storage on the first layer (2.4 GB instead of 4.38 GB), and recording will be possible on only one layer per side. Two-sided blank discs will probably be available to double the one-side capacity.

11.3 COMPUTER HARD DISKS

Video applications such as nonlinear editing rely heavily on the technology of the computer hard disk for their primary storage medium. PCs generally use their hard disk for storage of programs, data, audio, video, and so on, under the control of the file system of their operating system software. For random access purposes, hard disk tracks are divided into sectors in a manner similar to that used by CD-ROMs or DVD-ROMs. However, the file system manages the use of sectors quite differently.

The objective of a hard disk file system is to make available the full storage capacity of the disk for whatever information the PC user wishes to store. Separate segments of information are files, which the file system will place onto the disk according to its algorithm. This has several features:

- Files are given names and a directory is built at a specified location on the disk containing file names and their sector locations on the disk.
- Files can be added to or removed from the disk at any time.
- Files are assigned to sector locations in a way that assures that all the sectors of the disk can be used, regardless of file sizes or locations.

Notice that it is not usually an objective to require that all the information in a file is placed in physically adjacent sectors on the disk. That means there may be several head seeks required during the reading of a file. Since head seeking of a hard disk is relatively fast, this is not usually a problem with general computer data, but it can be a problem with time-sensitive data such as audio or video. Retrieval systems for these types of data must not depend on the continuity of data or its rate coming from the disk. Thus, buffers are usually set up for audio or video playback to smooth out the rate of information coming from the disk. In this case, the hard disk will provide a burst of data into the buffer, which is then read out at the desired constant playback rate. When the buffer content drops below a specified level, another burst of data is read from the disk.

High-performing video storage systems, such as those used with nonlinear editing, generally provide their own hardware and software for management of video playback (and recording), using the PC's main hard disks. Those disks must meet certain requirements for burst data rates, and the PC operating system must be able to give priority to the video read/write hardware when required during playback or recording. This latter requirement recognizes that other activities within the PC may also require hard disk access and these must be interleaved in the hard disk operation without causing video interruptions.

11.4 LASER VIDEODISC

The analog laser videodisc (LV) has been around for more than 20 years and has captured certain niche markets, especially in education and training. It was one of the first applications of laser optical recording and it delivers excellent SDTV picture quality. The system is available to users only in the form of a player device; recordings are always made at dedicated mastering houses that have the expensive facilities for recording and replicating. Until the arrival of DVD, LV had little competition in applications that demanded fast random access to prerecorded video clips or still frames. A significant infrastructure exists in the videodisc markets for the production of programs and discs; it will take some years for the same thing to occur for DVD.

Some of the characteristics and performance parameters for videodisc are shown in Table 11.2. The original LV standards provided two channels of FM audio frequency-multiplexed below the video FM spectrum; however, this gave limited performance and

Table 11.2
Characteristics of the Analog Laser Videodisc

Parameter	Value
Video standard	NTSC/PAL Composite
Disc diameter	30 cm
Disc thickness	2.4 mm
Disc rotational speed	600 to 1,800 rpm
Playing time/side	1 hr (CLV), 30 min (CAV)
Frame storage capacity (CAV)	54,000
Video modulation	Analog FM
Video bandwidth	5.5 MHz
Video SNR	50 dB
Audio channels	2 FM analog, 2 digital (optional)
Digital audio sampling	16 bps, 44.1 kHz
Compressed audio (Dolby Digital)	384 kbps
Digital audio dynamic range	96 dB (uncompressed)
Digital audio bandwidth	20 kHz

several digital audio options were later developed using either uncompressed PCM or Dolby Digital compression.

The large 30-cm disc size has fixed the size of LV hardware. However, when the CD was introduced, a 20-cm disc standard was introduced and most LV players now can play either disc size.

11.5 CONCLUSIONS

Disk storage of video has definitely come of age in both magnetic and optical technologies. With the higher recording densities of computer hard disks and DVD, and with digital video compression, single disks or drives can now store several hours of high quality video. The inherent advantage of disk storage for rapid random access to the information supports applications like nonlinear editing, which could never be done with tape media.

12

Trends in Recording

The development of video recording marked a turning point for the video industry—it was now possible to capture video at a different time than it would be displayed, and the captured video could be assembled to produce programs meeting all the technical and artistic objectives of professional producers. The present range of video applications could not exist without video recording.

In this final chapter, I will discuss the various technological ingredients supporting video recording with regard to future trends and possibilities. Since this is of necessity a personal opinion, I will speak mostly in the first person to distinguish from the rest of the book, which is more factual.

12.1 TECHNOLOGICAL TRENDS

Many diverse technologies contribute to video recording. These, shown in Figure 12.1, have themselves progressed to make possible the tremendous advancements in video recording equipment and systems since the first commercial unit was introduced in 1956.

12.1.1 Recording Area Density

The area density factor specifies how much information is recorded per unit area of the record medium (see Section 2.1.2). A curve for the growth of area density in magnetic recording since 1970 was shown in Figure 2.2. Figure 12.2 includes that information as well as a second curve for the growth of optical recording area density.

I believe this growth will continue into the future, since theoretical limits are not close. The limits are mostly practical ones of design and manufacture; these have the habit of being regularly overcome as development continues. An important factor determining how much development effort is applied is the size of the related markets; recording markets, both home and professional, are healthy and growing and can support large

Figure 12.1 *Technologies contributing to video recording.*

development programs for continuing long-term technology improvement. As video recording becomes digital, it also benefits from the computer industry's development efforts to improve storage and processing.

Many technologies contribute to the improvement of recording area density, ranging from the physical characteristics of recording materials, heads, and decks, to signal processing capabilities of integrated circuits, computers, and software. These are all discussed below.

Improvements in area density affect several aspects of recording systems, but they generally cannot be exploited without creating new standards. Because of the need for new standards, density improvements are generally taken in large steps rather than incrementally. The combination of picture quality, recording time, equipment size, and cost moves to a new level with higher density; size and cost may be reduced and recording time and picture quality may be increased—the improvement can be applied to any combination of these four attributes. For example, a new system may provide cost and size reduction with the same recording time and picture quality. Another combination could be to improve picture quality (e.g., HDTV), while keeping equipment size, playing time, and cost the same as previous standards.

A new standard requires a new infrastructure of equipment suppliers, media sources, and services. All parties (producers and viewers) who will use the new standard have to become equipped for it before the volume of use can grow. In some respects, it is a chicken-and-egg situation—people must make up-front investments before benefits can accrue.

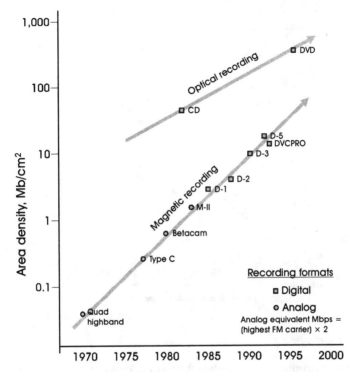

Figure 12.2 *Growth of the area density factor for magnetic and optical recording.*

Density improvements allow new recording applications. Magnetic recorders were once too large and costly to make practical hand-held camcorders; however, density increases overcame this limitation by making possible smaller tapes that reduced equipment size and weight but could still achieve the necessary picture quality and recording time. Similarly, density improvements made possible practical optical discs that could record several hours of video on a single disc.

12.1.2 Integrated Circuits

As with recording density, integrated circuits have undergone aggressive improvement throughout their history. This is generally described in terms of the number of devices on each IC chip, which was first articulated by Gordon Moore of Intel. In 1965, he predicted that IC devices per chip would increase by a factor of two every 2 years—a remarkable prediction, now known as *Moore's Law*, that is still true today. This is shown in Figure 12.3. Since the cost per chip increases very little with the complexity, this represents more processing power at nearly the same cost. Industry experts say that Moore's Law could go for another 5 to 10 years, which has massive implications for all electronics products, video recorders included.

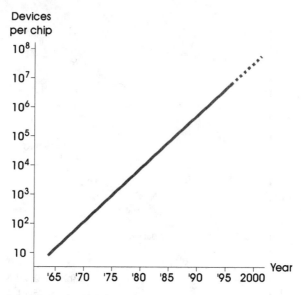

Figure 12.3 *Moore's law. Reproduced with permission from [1].*

12.1.2.1 Microprocessors

A key IC is the microprocessor, which is the heart of every PC and is also found embedded in most recording products. IC advances are converted into more powerful microprocessors at ever lower costs. In recorders, microprocessors not only perform system control functions, they are increasingly a part of the audio and video signal processing. So far, video signal processing with microprocessors has been limited by device speed, but with improvement of 2× every 2 years, speed will be less and less of a limit in the future.

An advantage of computed signal processing is the flexibility of having algorithms in software instead of being hard-wired—systems can contain several algorithms for different standards at little extra cost. Algorithms can be updated simply by updating the software; no hardware need be changed. My opinion is that in 5 years, most video processing for PCs will be in software. However, recorders cannot afford the overhead of a full PC, but they may use software processing with a video digital signal processor (DSP) chip, which is a microprocessor optimized for signal processing.

12.1.2.2 Application-Specific ICs

In situations where software-based processing is not appropriate, the efficiency of ICs is available with an *application-specific IC* (ASIC). These custom ICs are easily designed for any application with modern ASIC design tools that use the power of PCs.

One major area that has benefitted from ASICs is video compression. Inexpensive chips for various compression and decompression algorithms have made compression practical at all levels of recording. Cost is no longer a factor in deciding whether to use compression; the principal considerations now revolve around the performance trade-offs and the effects of compression on editing. The significance of ASICs is that new recording products can use the future advances in IC technology to improve performance while reducing size and cost.

12.1.2.3 IC Memory Devices

The capacity and cost of memory devices respond directly to the improvements in IC density. This has reached the point where megabytes of memory are practical in almost any type of product. Video frame memories cost only a few dollars; recorders can have as many of them as needed for very complex signal processing. The cost of a memory-based process such as a time base corrector has become trivial.

12.1.3 Mass Storage

Computer mass storage (fixed and removable disk storage) is also on a steady improvement curve. Powered by the advances in recording density, mass storage capacities and speeds are increasing while costs are decreasing. This will continue, leading to more video application of general-purpose computer mass storage. The principal field where this will be felt is nonlinear editing, which promises to become less costly while improving video quality and storage capacity.

Another important mass storage application is the video server. The objective of a server is to store a large amount of video, any part of which is available on demand to one or more output lines. This generally involves multiple large hard disk drives in an array capable of a high output data rate. The disk array data stream is partitioned to a number of output circuits to provide simultaneous multiple video streams originating from different locations in the storage.

So far, video use of computer mass storage devices has only been in fixed equipment. However, as the capacity grows in small mass storage devices, the potential increases for using these units in camcorders and other portable equipment. But, hard disk storage in a camcorder raises the problem of how the data is retrieved from the unit. The hard disk can be made removable, so it is physically taken to a separate playback unit for retrieval. The effectiveness of that is limited by the cost of hard disk drives—they are much more expensive than tapes and certainly cannot be exchanged in the same way as tapes. An alternative is for the camcorder to play back its video to a communication line or transmission system. This ties up the camcorder while it is being done, a problem that can be reduced if the hard disk can play out its data faster than real-time video playback, which is becoming possible.

Another possibility for a camcorder is to use an optical recording drive having removable media, such as DVD-RAM. This may soon become practical in portable equipment. Recorded DVD-RAM discs could be handled like tapes—in fact, they may well cost less in the long run.

12.1.4 Video Compression

For any application concerned with cost, digital video recording is not practical without some amount of compression. Existing standards, such as MPEG-1 and MPEG-2, offer the flexibility to adjust compression parameters to match application needs. In my opinion, however, there still remains much room for advancement in compression, both in the design of algorithms and in their application. As advancements occur, the trick will be in getting them into the standards without completely upsetting the market.

When digital video compression first appeared, professional video producers said they would never use it because it represented a trade-off in the picture quality margin that allows professional postproduction tasks to be done without degrading the picture. This is still true, but it doesn't take as much compression to record video with current recording densities, and modest amounts of compression that don't trade off too much quality margin can be used. A large part of the market has accepted this and the argument about whether to compress or not is slowly going away.

At the same time, further advances in area density will occur and the cost of recording uncompressed digital video will keep coming down. Soon, many users of SDTV will be able to afford uncompressed recording and the issue of compressing or not will move into the HDTV arena.

12.1.5 Optical Technology

Advances in optical components, lasers, and sensors made the CD mass market possible, and now further advances have spawned DVD. Will there be more? The only safe answer is "yes," because there's still a long way to go, especially with recordable and rerecordable media. Shorter wavelength lasers are in the pipeline and additional advances in density are out there. A density advance, however, means another new standard, and if the market embraces DVD the way everyone expects, no one will want a new standard for at least 5 years. But I think another factor of five or so increase in density will occur when the market is ready for it.

The improvement needed sooner is in recordable and rerecordable optical techniques. There has been fair market acceptance of CD-R, but CD-RW is mired in confusion. I think that will clear up, but maybe not before DVD-RAM appears on the market. The industry needs to settle on the best approach and move forward in a single direction.

12.1.6 Broadcast, Cable, and Satellite TV Distribution

SDTV distribution by broadcast, cable, and satellite has never been stronger. This, of course, represents a substantial recorder market that already is going entirely digital. At the same time, these customers are facing a transition to DTV and HDTV. In the United States, it is already under way. This transition is creating a market for multiple-standard recorders that will handle both SDTV and HDTV. This is entirely feasible with digital technology, and standards for this class will surely come.

12.1.7 Personal Computers

Advances in ICs (microprocessors and memory) have already been mentioned. These guarantee that the performance of PCs will continue to increase, making video operations by PCs more powerful and cost-effective. This applies to all markets, both professional and home.

When a user already has a standard PC, it is advantageous to use that for video processing, if possible. However, it is not generally practical to embed a standard PC in a recorder for signal processing purposes; a DSP chip is a better answer for a programmable processor in a recorder. The principal use for a standard PC in recording is as an editing controller. In that case, a standard PC's price-performance ratio would be hard to beat with any custom-designed processor.

An important controversy in the industry is whether the PC or the TV will become the center of home electronics. In my opinion, this is moot, because the PC and the TV are merging—PCs are showing video and TVs are embracing computers. At some point in the future, maybe 5 or more years, there will likely be a PC-TV home product (or a TV-PC—let's call it a home electronics center) taking a significant part of both markets. However, the marketing success of this in the home will depend very much on overcoming the consumer prejudice against PCs as being too difficult to operate. Both industries are working on the problem of difficult-to-operate PCs—the PC industry is addressing it directly, and the consumer products industry is moving to include a PC in a TV but make the consumer not see it as a PC. Success of either approach would create a large market.

12.1.8 Production and Postproduction

TV programs, training videos, and semiprofessional videos all now are created using the production-postproduction method (see Chapter 8). A large part of this is done with nonlinear editing, although there is still a lot of industry effort to make linear (tape) editing easier to use. I think the continued cost reduction and performance improvement of nonlinear systems will eventually take over the whole market—even the home.

Videotape editing for the home market is so awkward and difficult that few home users go through the trouble. However, with PC-based editing and more powerful PCs in the home, nonlinear editing will receive more interest. When a home user can connect a digital interface cable between his camcorder and his PC and have the PC capture the video faster than real time, one big limitation to home nonlinear editing will be overcome. Easy-to-use editor software will be available at low cost, and home PCs will be able to output high-quality video back to the camcorder or simply display it from the hard disk directly to TV monitors.

12.1.9 Mechanical Manufacturing

Video recorders are electromechanical devices that depend heavily on the quality of their mechanical manufacturing. Tape decks, disk drives, magnetic scanners and heads, and optical heads are all basically mechanical assemblies that have to be mass produced to very tight tolerances and high quality, yet at low cost. This requires very sophisticated manufacturing technology; its mastery is essential to the success of a recording product.

It has been shown again and again that proper design of mechanical parts and their manufacturing process can achieve remarkable precision and quality at low cost. A good process that produces the desired result almost all the time reduces costs associated with inspection, rejects, and repair. Thus, quality costs less when it comes automatically out of the process itself.

Manufacturing of recorder parts, such as heads, often involves very small dimensions with tight tolerances. These cannot be built efficiently by normal mechanical fabrication methods such as casting, machining, and so forth. More and more, heads are being designed to use chemical processing techniques developed for IC production, such as deposition, sputtering, etching, plating, and so on. These methods lend themselves to processing a number of parts at the same time (*batch processing*), which reduces cost and, with good process controls, can produce high quality.

12.1.10 Servomechanisms

All recorders have servomechanisms for control of tape speed, disk rotation, tracking, or other parameters. Servos are a mature technology whose effectiveness depends largely on the quality of the components in the loop—sensors, amplifiers, and motors or actuators. Incremental improvements will continue to happen, but this area primarily will follow the needs posed by the rest of the recording system.

Video recording is one of the largest markets for servo components, which has spurred a lot of development for reducing cost and increasing quality of these parts.

12.1.11 Materials Science

Many advances in recording are the result of new or improved materials. Magnetic material, optical materials, substrates, and head parts all have unique material requirements that often are not common with other fields. Existing materials for recorders are the result of materials research directed at the needs of recording. This is continuing and can be expected to contribute to further improvement in recording density, media life, and cost.

New systems depend on the creation of appropriate materials to build recording media and recorder parts. Close cooperation between materials researchers and system designers is necessary to achieve the best performance that current materials technology can deliver.

12.1.12 Digital Transmission

Recorders themselves do not involve transmission (unless you think of the recording process itself as transmission), but the systems in which they are used do have transmission. Some of the interconnections used with recorders were discussed in Section 9.5. Longer distance interfaces, however, are sometimes used and are affecting recorder design. This is especially important for transmission of production material from the field back to a studio or production house.

When digital video is being transmitted for storage at the receiving location, there is no reason to stick to normal playing rates. Recorders can be designed to run faster than normal play speed for data transfer over wideband links such as satellites or high-speed computer networks. A speed race similar to that which occurred with CD-ROM drives (see Section 11.1.1.3) will occur here. Higher-than-normal-speed transmission is an important new area of recorder application.

12.2 RECORDERS

Now, I will take all the points raised in the previous discussion of underlying technologies and project what this may mean to recorders in both professional and home markets. In each section, I will first discuss recorder applications and what the markets are looking for; then I will give my opinion of new product developments. My time frame for this is about 5 years; going beyond that is really blue sky.

12.2.1 Professional Recorders

These recorders include broadcast, cable, production and postproduction, and other high quality video delivery systems. Both SDTV and HDTV recorders are discussed.

12.2.1.1 Applications

Professional recorders are used in program creation systems (production and postproduction) and in real-time program delivery systems for broadcast, satellite, and cable distribution. Program creation includes not only that for conventional TV programs, but programs for transfer to motion picture film, training programs, programs for distribution via the Internet, and semi-professional creation of programs for weddings and other special events. Recorder objectives for professional applications generally place performance, flexibility, and reliability first, with cost being subordinate.

An important application area is the *video server* for real-time distribution of video on demand, editing, and broadcast, cable, or satellite distribution. In these cases, banks of recording devices store video for random access playback to one or more output lines. This is a rapidly growing market.

12.2.1.2 Products

Standards—New recording formats have to become formal standards before they can attract much market interest. With the new market needs of HDTV recording, and area density improvements that come out of the R & D pipeline, there are sure to be some new standards in the next 5 years. Otherwise, the existing standards for SDTV digital recording will continue to be viable. The most development will probably occur in the small tape formats such as the 6.35 mm (1/4-in) size.

Editing—Professional editing systems are available from very simple to highly sophisticated systems. As equipment costs come down, there will be more use of high resolution scanning formats carried through production and postproduction and archiving, with conversion to distribution formats only at the end of the process. Thus, program material can easily be distributed in any format where there is a market.

Lower equipment costs, smaller recorders, and more computer power for the money will increase the use of field editing equipment. Program producers can make preliminary edit decisions in the field while retakes are still possible. This will reduce the incidence of problems that must be solved in postproduction.

Camcorders—Many camcorder improvements will be to the camera portion, which will not be discussed here. Regarding the recorder part of camcorders, smaller tape formats resulting from recording density improvements will be significant. In addition, hard disks or DVD-RAM for recording in camcorders will be introduced. In addition, faster data transfer from camcorder to editor will be a factor in most professional camcorders.

Optical recorders—As higher data rate optical recorders such as DVD become available in recordable formats, there will be more use of optical for professional archiving of programs. However, the DVD data rate requires a lot of compression even for SDTV, and there would be opportunity for a higher-rate system to reduce the amount of compression (and thus degradation) that is needed. There are such systems on the market now, but they have not been widely accepted because they are single-supplier products that have uncertain long term support. Products based on DVD and supported by the DVD community would likely have more acceptance.

12.2.2 Home Recorders

Products for the home market generally place the highest priority on cost—if the cost is not low enough, there is no market. But once cost has been considered, the market is very demanding of reliability, performance, and flexibility, in that order.

12.2.2.1 Applications

Home recorder applications fall into three categories: playback of prerecorded videos, rented or purchased; recording and playback of video taken from broadcast, cable, or satellite sources (with proper consideration of copyright regulations); and production of home programs for family archiving and pleasure. These have very different recording requirements.

Prerecorded video playback—These units generally are designed as set-top boxes and interface with a TV for display. VHS tape recorders and DVD players are made that way. Of course, VHS recorders will be used for prerecorded video for many years to come, simply because of the investment in recorders and tapes. It will take quite a few years for DVD to build an equivalent infrastructure, although that is an essential ingredient of success.

The current DVD players are at a disadvantage because they do not record and thus, they cannot support the other two classes of home application. DVD-RAM recorders will overcome that limitation, but they will take some years to reach a competitive price point to compete with the recording capability of VHS. The DVD community has a lot of work to do.

Recording "off the air"—This is a capability of VHS recorders that is used extensively by a small segment of the market—those people who have the inclination and will take the trouble to record programs. Many VHS owners never record and use the machine only to play prerecorded tapes. It would seem that the digital tape formats developed for camcorders would be valuable for off-air recording, but there are no digital set-top tape decks yet. Once DVD-RAM is available, it would also fill this need. If that happens soon enough, a digital set-top tape recorder may never occur.

Home program production—The heart of home program production is the camcorder. These are available in VHS and 8-mm analog formats, and the DV digital format. As the DV equipment comes down in price, it will begin to command more of the market because its picture performance can deliver excellent edited copies. One of the reasons that home editing has been little used is that the picture quality of VHS or 8-mm second generation is so marginal.

Editing equipment is an important part of home program production. The camcorder can be the signal source, but another recorder is needed to make edits. I think this will move quite rapidly to digital nonlinear editing using the home PC; this will especially happen after DV camcorders start selling to the home. Of course, edited programs could be played directly from the PC for viewing, but this will rapidly use up even the largest PC hard disks. The answer will be DVD-RAM; completed programs can be off-loaded to DVD-RAM disks, which can then be viewed using the home's DVD player.

12.2.2.2 Product Summary for the Home Market

Camcorders—These will go to DV over the next several years as the market volume increases and prices drop. Compatible nonlinear editing hardware and software for use with a home PC will support the users who wish to create edited programs. I think this will increase as systems become easier to use and consumers see the results that can be produced.

Set-top recorders—The future set-top recorder will be the DVD-RAM. I doubt that a new tape recorder for the set-top will succeed, even with DV. The advantage of DVD-RAM is that it also can play prerecorded DVD discs, which are going to be available long before any new tape format could come on line. Thus, one set-top box supports all applications.

Players—The home player for prerecorded material will be the DVD player. These units will fall in price to the level of VHS recorders. As digital TVs start selling, a digital interface between the DVD player and the TV will become necessary. Then, the full performance potential of the component recording in DVD will be realized.

Home networks—Much industry attention has been given to designing an audio-video network for the home that would allow video players and recorders to be centralized, with their services available to multiple displays around the home. The reception of incoming signals from broadcast, cable, satellite, or the Internet could also be centralized in the network scenario. This has many attractive features, but it is unknown whether it will ever get off the ground. I expressed my opinions about this in another book [2].

12.3 CONCLUSION

Video recording markets are healthy and growing. They are large enough to support the range of research and development needed to keep the technology advancing as fast as the market can absorb it. The technological opportunities abound for improvement of product performance and reduction of cost. New recording applications, such as DVD, video servers, digital broadcasting, and so on are increasing the availability to the public of recorded video programs.

REFERENCES

[1]   Luther, A. C., *Video Camera Technology*, Artech House, Inc., Norwood, MA, 1998.

[2]   Inglis, A. F., and A. C. Luther, *Video Engineering,* 2nd Ed., McGraw-Hill, New York, 1996, Ch. 19.

Glossary

4:2:0 In a digital component system, where color-difference components are subsampled 2:1 both vertically and horizontally.

4:2:2 In a digital component system, where color-difference components are subsampled 2:1 horizontally only.

4:4:4 In a digital component system, where there are no subsampled components. This format can be RGB or YC_RC_B.

A-B roll editing In videotape editing, two source recorders play prepared tapes containing alternate scenes of the final program. The signals go to a video switcher under computer control; the switcher output goes to a third recorder that records an edited master tape.

AC-3 See Dolby Digital.

acicular Rod-shaped. Used to describe the magnetic particles in the coating of gamma ferric oxide magnetic tape.

active line time In video scanning, the time during line scanning where picture information is actually transmitted.

active scan interval In a scanning system, the group of pixels that represents useful video information. The rest of the pixels are usually black and are part of the blanking intervals.

ADC See analog-to-digital conversion.

additive color A color reproduction system that reproduces colors by combining color light sources of the primary colors. See also subtractive color.

adaptive differential PCM (ADPCM) A differential PCM scheme where the scale of the difference values is changed dynamically according to the signal properties.

ADPCM See adaptive differential PCM.

Advanced Television Systems Committee (ATSC) An organization in the United States having the objective of standardizing advanced television systems, including the new digital television system adopted by the Federal Communications Commission.

algorithm The definition of methods used in a specific process, such as an algorithm for video compression.

aliasing In a sampled system, distortion caused by sampling input frequencies above the Nyquist limit.

alpha-wrap In a helical scanning videotape system, a design where the tape wraps completely around the scanner. See omega-wrap.

analog A system that represents signal quantities on a continuous scale.

analog-to-digital conversion (ADC) The process of converting an analog signal to a digital representation. It includes the steps of sampling, quantizing, and encoding.

application-specific IC (ASIC) An integrated circuit (IC) designed for a specific task, such as camera signal processing.

ASIC See application-specific IC.

aspect ratio The ratio of the width to the height of a picture.

assemble editing Videotape editing where new recordings are placed at the end of previous recordings.

astigmatism An optical distortion where the focused spot is elongated and never focuses to a sharp, round spot.

asymmetrical compression A compression system where different amounts of processing are required for encoding and decoding. Many systems have a complex encoding process but their decoding process is simplified to reduce cost at the receiving device.

ATSC See Advanced Television Systems Committee.

audio-CD The original Compact Disc, designed for distribution of prerecorded audio to the home and defined in the *Red Book*.

azimuth In a magnetic head, the angle that the head gap makes with the track.

azimuth recording In helical scanning, the method of offsetting the azimuth angle between adjacent tracks to reduce crosstalk when the tracks are placed close together without guard bands.

bandwidth The amount of frequency spectrum required by a system. Bandwidth is often associated with a baseband system, where it equates to the highest frequency component contained in the signal.

baseband system A system whose bandwidth begins at a low frequency (e.g., less than 60 Hz or even zero frequency).

BER See bit error ratio.

bias recording In magnetic recording, the use of a high-frequency signal combined with an analog signal during recording to linearize the magnetic transfer characteristic for the analog signal.

bidirectionally predicted frames In a video compression system, frame reproduction based on both the previous and the next frame of the video sequence. Bidirectional frames (B-frames) provide the most compression. See predicted frames and intracoded frames.

bimorph A structure made of two layers of dissimilar material. As used in recording, the materials are a piezoelectric ceramic on a metal substrate. When a voltage is applied to the piezo material, the structure bends. This is used for transverse movement of heads on helical scanners.

binary In a digital system where the symbols have only two values—one or zero. A binary symbol is called a bit.

birefringence The property of an optical medium to display a different velocity of propagation for different linear polarizations of light.

bit See binary.

bit error ratio (BER) The number of bit errors in a given number of bits transmitted, specified as the ratio between the number of errors and the total number of bits. BER is usually expressed as a negative power of 10; for example, one error in a million bits is a BER of 10^{-6}.

bits per sample In a sampling system, the number of bits used to encode each sample. Abbreviated "bps."

black level In a video system, the signal level or value representing the darkest (blackest) areas of the picture.

blanking interval In a video system, the portions of time in the signal allowed for horizontal and vertical flyback of the display. Analog video systems always have blanking intervals, but digital video systems may or may not.

brightness To a human observer, is the sensation of intensity in a picture. See luminance and saturation.

burned-in time code Time code that is displayed in the video picture. This is often done during editing, but removed from the final copies.

burst errors In a digital system, errors in a group of sequential channel symbols.

camcorder A unit that combines a camera and a video recorder in one package. Camcorders are generally meant to be portable and carried or held by one person.

camera A unit that views a scene and captures a picture of it. Cameras may be based on photographic film, or they may be electronic. They may capture stills or motion sequences. Photographic cameras inherently store their pictures on the film, but electronic cameras may deliver a signal into a system and/or store pictures on an electronic recording device.

capstan drive In a tape deck, the use of a rotating shaft pressed against the tape to control the tape speed. Pressure of the tape against the capstan is often provided by a pinch roller.

cassette A holder for tape. It usually consists of two reels with the tape captive between them, and one or more openings where the tape and reels can be mechanically accessed by a tape deck for threading.

CAV See constant angular velocity.

CD See Compact Disc.

CD-audio See audio-CD.

CD-R See recordable CD.

CD-ROM The computer data version of the Compact Disc. It is a read-only medium.

CD-RW See rewritable CD.

channel The signal path for recording or transmission.

channel signal In a recording or transmission system, the signal that passes through the physical channel. It normally uses some type of encoding.

chrominance In a composite video system, the signal that transmits the color-difference components.

clamping In analog video systems, a process of setting some portion of the signal to a fixed level to restore the DC component. Usually a part of the horizontal blanking interval is devoted to clamping.

clock signal In a digital system, a signal that indicates the timing position of the bits for the purpose of data separation.

CLV See constant linear velocity.

CMYK A system for color printing that uses the subtractive color primaries cyan, magenta, and yellow plus black.

coaxial cable A two-conductor cable that has a central conductor surrounded by a uniform-thickness layer of insulation, which is covered by a coaxial sheath that forms the second conductor. The central conductor is used for the signal and the coaxial sheath forms the return or ground conductor.

coercivity In magnetic materials, the magnetizing force needed to reverse the remanent magnetization to zero.

color-difference signals In a component color video system, the two signals that represent the color and color intensity of the scene. Color-difference signals go to zero for white regions of the scene. See YC_RC_B format.

color subcarrier A high frequency carrier modulated by one or both color-difference signals and multiplexed into a video channel along with the luminance signal. Color subcarriers are used in the NTSC, PAL, and SECAM video systems.

color-under A system used in analog home video recorders where the chrominance information is recorded in the same track but separate from the luminance information by placing it in a frequency range below the spectrum of the frequency-modulated luminance information. This method allows a very inexpensive means for time base correcting the color information but not the luminance information.

Compact Disc (CD) A record medium in the form of a 12-cm plastic disc that is read by an optical sensor. The first CD was developed for distribution of prerecorded audio (CD-audio) and was introduced to the home market in 1982. It uses PCM digital encoding of stereo audio with a playing time up to 74 min. Subsequent developments have produced the CD-ROM for use with computer data, the CD-R recordable CD, and others. Note that the unusual spelling of "disc" when referring to the CD is part of the trade name.

complementary colors In a color reproduction system, colors produced by combining the primary colors two at a time. For example, in an RGB system, the complementary colors are magenta (red-blue), cyan (blue-green), and yellow (red-green).

component video A video system that separately processes and transmits the three primary color channels. See RGB and YC_RC_B format.

composite video An analog video system where the three primary color channels have been combined in some way to send them through a single transmission channel. See NTSC, PAL, or SECAM.

compression In digital systems, any process that reduces the amount of data needed to transmit information. See lossless compression and lossy compression.

constant angular velocity (CAV) In a disk drive, a constant disk rotational speed, which gives a varying track speed according to the track radius.

constant linear velocity (CLV) In a disk drive, control of the disk rotation to give a constant track speed independent of track radius.

contouring A picture distortion caused by having too few quantizing levels for the luminance or color-difference information. It gives an effect similar to the contours on a geographical map.

Curie temperature In a magnetic material, the temperature above which most magnetic properties are lost.

cuts In editing, abrupt transitions between one shot and the next. Cuts normally occur during the vertical blanking interval.

DAC See digital-to-analog conversion.

data density ratio In an encoded digital transmission or recording system, the ratio between the minimal time between transitions of the channel signal and the minimal time between transitions of the incoming data stream before encoding. A higher density ratio indicates that more information is transmitted for a given channel bandwidth.

data rate In a digital system, the transmission rate of bits, usually given in bits/s (bps), kilobits/s (kbps), or megabits/s (Mbps).

data separation The process of extracting the encoded bits from the digital channel signal.

DCT See discrete cosine transform.

decoding In an encoded system, the process of undoing the encoding and recovering the reproduced signal.

differential PCM (DPCM) A PCM system where the transmitted signal represents the differences between the incoming words or samples. See predictive coding.

digital A system where signals are represented by symbols having a discrete scale of values. For example, symbols in a decimal digital system have 10 possible values or in a binary digital system they have only two values.

digital component video A digital video system where the video components are separately digitized. See component video.

digital filter In a sampled digital system, a process that manipulates the sample values so as to modify the frequency domain of the information content of the samples. Typically this is done by accumulating samples that have been delayed and weighted by different values. See finite impulse response filter.

digital television (DTV) A television system where the transmitted signal is digitally encoded. See encoding.

Digital Versatile Disc (DVD) A new digital optical disc format for prerecorded audio, video, and computer use. It is the next generation of the Compact Disc. DVD-ROM and DVD-RAM (recordable) versions will also be available.

digital video Any video system where the signals are digitally encoded.

Digital Video Broadcasting (DVB) A standard for digital television developed in Europe. The DVB standard is used worldwide for satellite broadcasting of television.

Digital Video Cassette (DVC) Now called DV. A digital videotape system developed for home camcorder use. It uses a small cassette holding 1/4-in tape.

digital-to-analog conversion (DAC) The process of converting a digital signal back to analog representation.

digitize A synonym for analog-to-digital conversion.

digitized composite video The digitizing of analog composite formats such as NTSC or PAL. This is often advantageous when digital equipment is used in an otherwise analog video system.

discrete cosine transform (DCT) In video compression, a form of encoding that transforms a two-dimensional array of pixels (usually 8×8) to a two-dimensional array of frequency coefficients. This transformation facilitates compression because many of the frequency coefficients have zero or small values and can be ignored or quantized with fewer bits. Statistical coding of the frequency coefficient values provides further compression. See quantizing.

disk drive A mechanism for recording or replaying tracks on a disk. It consists of a motor that rotates the disk and a record/replay head mounted on a mechanism that moves the head in a radial direction across the disk surface. Some disk drives have multiple disks and heads to increase storage capacity.

dissolves In editing, transition effects where the video of one shot fades out while the video of the next shot fades in.

dockable In camcorders, a packaging approach that allows camera and recorder to be separated. This is useful when the camera portion is used without a recorder in a studio setting or when the same camera is to be used with different recorders.

Dolby Digital A multichannel audio compression technique, used in the U.S. ATSC digital television standards. Also known as AC-3.

DTV See digital television.

DV See Digital Video Cassette.

DVB See Digital Video Broadcasting.

DVC See Digital Video Cassette.

DVCAM A professional recording system developed by Sony. It is based on the Digital Video Cassette home recording system.

DVD See Digital Versatile Disc.

dynamic range In a signal system, the ratio between the largest and the smallest signals that can be handled.

E-E See electronics-to-electronics.

eddy current losses Resistive losses in a magnetic circuit caused by induced currents in the magnetic materials.

edit decision list (EDL) A list of time code addresses and commands for control of an editing session.

edited master A recording made directly from an editing session.

editing In postproduction, the process of assembling program segments cut from raw shots into a finished program sequence. Editing may also include special effects to provide enhanced transitions between program segments.

EDL See edit decision list.

EFM See eight-to-fourteen modulation.

eight-to-fourteen modulation A method of digital encoding where 8 incoming bits are encoded as selected values of a 14-bit code.

electronic news gathering (ENG) The use of a camcorder for capturing news events.

electronics-to-electronics (E-E) In a recording system, the record electronics is connected directly to the playback electronics without going through record and playback. This may be used during recording to view the input signal by means of the playback monitoring equipment. It is also used for testing the electronics.

encoding The process of representing information by an electronic signal. This may be either analog or digital. Digital methods provide many more options for encoding, including compression techniques that reduce the required data rate and bandwidth for transmission.

ENG See electronic news gathering.

erasing The process of restoring a record medium to its unrecorded state.

ergonomics The study of human-equipment interaction. It is often called human engineering.

error protection In a digital system, the technique of inserting redundancy information before recording or transmission, so that errors may be detected and possibly corrected at the receiving end. See parity, Reed-Solomon code.

eye pattern An oscilloscopic display of a digital channel signal, synchronized to the recovered clock signal. It shows the quality of the data separation process.

fades In editing, transition effects where the video of one shot fades out and then the video of the next shot fades in.

FDM See frequency-division multiplexing.

fiber-optic cable A cable containing one or more optical fibers for signal transmission. It is being used more widely for digital signal transmission.

field In interlaced scanning, one vertical scan period, which is the scanning of all odd lines or all even lines. Two fields are required to make a complete frame.

finite impulse response filter (FIR) A form of digital filter that synthesizes its output response by summing a finite number of delayed and weighted copies of the input signal.

FIR See finite impulse response filter.

flicker The psychophysical effect perceived by the eye when viewing an image that is periodically refreshed at too low a rate.

floppy disk A disk medium using a flexible substrate.

flyback In a scanned video display, the moving of the scanning spot from the end of one line to the start of the next line, or from the end of one field to the start of the next field.

flying erase heads Heads mounted on a helical scanning drum for the purpose of selectively erasing helical tracks.

frame In a scanned video image, the complete scanning pattern including all lines and pixels.

frame rate The rate at which frames are scanned.

frequency interleaving In composite video systems, the choice of the color subcarrier frequency so that sideband components of the chrominance signal fall between the sidebands of the luminance signal. This reduces interference between the components.

frequency modulation An analog encoding technique that conveys the information signal in terms of the frequency of the channel signal.

frequency-division multiplexing (FDM) Multiplexing of two or more signals in the same transmission channel by having them occupy different frequency bands.

full duplex A communication system that handles communication in both directions simultaneously. An example is the conventional voice telephone system.

gamma ferric oxide A particulate magnetic material often used in magnetic coatings on tape or disks.

gap See head gap.

generation In recording, a single pass through recording and playback. In videotape editing, multiple generations occur as the signal is repeatedly copied. In analog systems, this causes accumulated degradation of the signals. This may be avoided in some digital systems.

gray scale In a video system, the brightness scale that is displayed. The term is often used to refer to the shape of the amplitude transfer characteristic even for signals that do not represent shades of gray.

guard band In recording, space allowed between adjacent tracks to prevent crosstalk. Guard bands can be eliminated in magnetic helical scanning by the method of azimuth recording.

hard disk A digital storage device, generally consisting of a rigid magnetic-coated disk written and read by magnetic heads configured to "fly" on an air film above the rotating disk. See disk drive.

head In a recording system, the component that records on the medium or picks up signals from the medium.

head gap In a magnetic head, the nonmagnetic spacer that breaks the magnetic circuit to produce a fringing field for recording or replay.

HDTV See high-definition television.

helical scanning In videotape recording, a method to achieve a high writing speed by mounting one or more magnetic heads to a rotating drum. The tape is wrapped around the drum so as to produce tracks at an angle to the edge of the tape. See alpha-wrap, omega-wrap.

high-definition television (HDTV) A new-generation television system that has up to six times the resolution of SDTV systems and a wider aspect ratio screen. It is a feature of DTV systems.

highlight In a video image, the response produced by an excessively bright spot in the scene. Highlights represent the highest video amplitudes and often have to be compressed at the camera to reduce the dynamic range of the signal to a displayable value.

horizontal blanking interval In an analog TV system, the period of time allowed for the receiver scanning beam to retrace back to the start of the next scanning line.

horizontal resolution In a video system, the degree of reproduction of fine detail of vertically oriented objects in the scene. It is quantified in TVL.

HSB See hue-saturation-brightness.

hue The attribute of color. It is the answer to the question, "What color is it?"

hue-saturation-brightness (HSB) A system of color measurement that specifies the hue, saturation, and brightness of the object.

Huffman coding A technique of encoding that assigns bits according to the frequency of occurrence of values. For example, a high frequency-of-occurrence value is encoded with a small number of bits, whereas values that occur less often are encoded with words of more bits.

human engineering See ergonomics.

hysteresis The property of magnetic materials to remember past conditions. Particularly valuable is the property to retain magnetization when the magnetizing force is removed. This is the basis for magnetic recording.

IEC See International Electrotechnical Commission.

IEEE 1394 A standard for serial digital communication developed for video equipment. See serial process.

insert editing In videotape editing, the new signal is recorded over a portion of an existing recording.

interframe coding In video compression, where the frames depend on each other. A common interframe coding technique is motion compensation.

interlaced scanning In a video system, the process of scanning half the lines in each of two vertical scans (fields). This is accomplished by having an odd number of total lines and making the ratio of horizontal to vertical scan frequencies equal to one-half the total line number. The result of interlacing is that large areas of the picture appear to be refreshed at the field frequency, which reduces flickering of the picture while allowing reduction of the frame rate.

interleaving In a digital system, a process where the data order is rearranged. This is usually done to break up burst errors so that the error protection system can correct them.

interline flicker An artifact of interlaced scanning where adjacent lines containing different horizontally oriented information flicker at the frame rate.

International Electrotechnical Commission (IEC) An international standardizing organization.

International Standardizing Organization (ISO) An international standardizing organization.

International Telecommunications Union (ITU) An international standardizing organization. Two branches of the ITU are relevant to video: ITU-R for radio broadcasting, and ITU-T for other forms of transmission.

intraframe coding In video compression, when frames are encoded independently of each other. This allows compressed video to be started or edited at any frame location. See encoding, interframe coding, predicted frames.

inverse DCT The process for decoding the discrete cosine transform.

ISO See International Standardizing Organization.

ITU See International Telecommunications Union.

Joint Photographic Experts Group (JPEG) A subcommittee of the ISO/IEC that developed an image compression standard for still images called JPEG compression. The standard provides a range of algorithms for compression of still images in both lossless and lossy modes.

JPEG See Joint Photographic Experts Group.

JPEG compression The name given to the image compression standard developed by the Joint Photographic Experts Group.

lapping A mechanical fabrication process where material is removed by abrasion. It is often used in magnetic head manufacturing.

line In a scanning system, the high-speed component of the two scanning motions. In most systems, the lines scan horizontally across the picture and field or frame scanning goes vertically.

linear time code (LTC) In magnetic tape systems, time code that is recorded in a longitudinal track.

longitudinal tracks In magnetic tape systems, tracks that run parallel to the tape edges.

look-up table In digital systems, the use of a memory to hold an array of values that are accessed by using the incoming signal values to address the memory. Thus, the incoming signal values are transformed to the set of values held in the memory.

lossless compression A data compression system that does not in any way change the information content of the data as a result of compression and decompression. Such a compressor can operate on any data without damaging it. However, the degree of compression is limited.

lossy compression A data compression system that may make changes to the information content of the data as a result of compression and decompression but it does that in a way that will have minimal effect on the usability of the data. This type of compression requires that the compressor know the data format and its use but, since that is known, much more compression can often be achieved than with lossless compression.

LTC See linear time code.

luminance In video systems, a signal that represents the visual brightness of the scene separate from its color properties.

magnetic heads In magnetic recording, the device that provides the recording magnetic field or, for replay, the device that picks up the field from the recorded surface and converts it to an electrical signal.

magnetic induction See magnetization.

magnetization The magnetic flux produced in a magnetic material by a magnetizing force.

magnetizing force In magnetic circuits, the magnetizing field strength produced by electric currents flowing in components such as wires or coils.

magneto-optical recording A recording method that uses a laser beam to locally raise the temperature of a magnetic medium above its Curie temperature to accomplish recording.

magnetoresistive heads In magnetic recording, playback heads that make use of the magnetoresistive effect, which causes the head resistance to change with magnetic field strength.

Manchester code A method of digital encoding that has a transition for every bit; a zero has a positive transition and a one has a negative transition. When consecutive values are the same, an extra opposite-direction transition is added between bits. Also called phase encoding.

mass storage In a computer, hardware for large-capacity nonvolatile storage of data. Typical mass storage devices are hard disks, floppy disks, CD-ROM, or magnetic tape.

modulation In a transmission system, the process that makes the information signal suitable to the channel characteristics. See encoding.

Moore's law In solid-state technology, an empirical statement first articulated by Gordon Moore of Intel that says: "The number of devices on an integrated circuit chip will double every two years." This has proven accurate since 1965 and shows no sign of reaching an asymptote any time soon.

motion compensation In video compression, a technique that examines sequential frames and determines the parts of the scene that have changed from one frame to the next. Only the new parts of the scene are transmitted to construct each new frame from previous frames. This is predicted frame compression. See also interframe coding.

motion vector In motion compensation, a signal that defines how a block of pixels in a new frame can be derived from a block at a different location in a previous frame.

motion-JPEG A video compression method that applies the JPEG compression standard to each frame of a motion video sequence. This is an example of intraframe coding.

Moving Picture Experts Group (MPEG) A working group of the ISO/IEC charged with standardization of motion video compression techniques. They are responsible for the MPEG compression standards.

MPEG See Moving Picture Experts Group.

MPEG compression A series of audio and video compression standards developed by the Moving Picture Experts Group. The standards use discrete cosine transform compression, statistical coding, and motion compensation.

NA See numerical aperture.

National Television Systems Committee (NTSC) The organization in the United States that developed that country's color television standard (called NTSC color television) in 1953.

noise In electrical systems, spurious or undesired components that are added to signals being processed or transmitted. Unavoidable random noise sources exist in most systems and place an upper limit on signal-to-noise ratio performance.

nonlinear editing Audio or video editing that uses random-access storage (usually a digital hard disk) to assemble edits in real time.

nonreturn to zero (NRZ) A digital encoding technique that encodes a binary one as one amplitude level and a zero as a different level.

NRZ See nonreturn to zero.

NTSC See National Television Systems Committee.

numerical aperture (NA) In a lens, a parameter that determines the diameter of the spot produced by focusing a laser beam. NA is equal to the sine of the half-angle of the cone of light passing through the lens.

Nyquist criterion In a sampling system, the fact that the sampling rate must be at least twice the highest signal frequency to avoid aliasing distortion.

Nyquist limit In a sampling system, the highest signal frequency that can be sampled without aliasing. See Nyquist criterion.

omega-wrap In a helical-scan videotape recorder, a design where the tape wraps part way around the scanning drum.

optical fiber An optical transmission medium comprising a microscopic glass fiber. Light energy sent into the fiber is transmitted within the fiber by total reflection at the fiber's surface. Optical fibers provide long-distance transmission with wide bandwidth.

outside broadcast (OB) The use of video equipment for on-location production.

packetizing In a digital system, the technique of breaking the incoming data into a series of blocks (packets). Packets may either be fixed in size or variable-sized, and each packet is handled separately by the system.

PAL See Phase-Alternating Line.

parallel process In an electronic system, where portions of the signal are processed simultaneously in separate channels. In digital transmission, where multiple bits are transmitted simultaneously over multiple circuits.

parity In digital recording or transmission, the technique of adding an extra (parity) bit to each word to make the bit values of the word plus the parity bit have an even sum (even parity) or an odd sum (odd parity). By testing the parity of each word at the receiving end, single-bit errors can be detected. This is a method of error protection.

PCM See pulse code modulation.

PCMCIA An interface standard for peripheral devices in portable PCs. The acronym stands for Personal Computer Memory Card Industry Association.

permeability In magnetic materials, the property that specifies the relationship between magnetic induction and magnetizing force. Permeability is often specified as relative, where it is the dimensionless ratio between the permeability of the material at hand and the permeability of a nonmagnetic medium such as free space.

phase encoding See Manchester code.

Phase-Alternating Line (PAL) A composite analog video television standard developed in Europe and now used by many countries around the world. PAL is based on most of the concepts of NTSC but has some improvements. The name comes from the line-to-line alternation of the color subcarrier phase that causes some types of transmission distortions to cancel. See composite video.

phase-lock loop (PLL) An electronic circuit that synchronizes a variable local oscillator to an incoming signal. A PLL is often used for clock signal generation in data separation.

picture-in shuttle In videotape recording, the ability of a system to display picture signals while running the tape at variable speeds. See shuttle mode.

pinch roller In a tape deck, a compliant roller that presses the tape against the capstan to allow the capstan to control the tape speed.

pits and lands In optical recording, the term often used to describe the recorded and unrecorded areas of the tracks.

pixel In video systems, an independent point in the picture, separately definable for brightness and color. Video systems usually have a two-dimensional array of pixels; there may be hundreds of thousands or more pixels in a single frame.

platter In a disk recording, a name for the disk substrate. Usually used to refer to rigid substrates. See hard disk.

PLL See phase-lock loop.

postproduction In the style of program creation where all shooting (production) is done before editing, the process of program editing, assembly, and special effects.

pre-roll In tape editing, the technique of starting tape decks a certain amount of time ahead of the edit to allow time for servos to synchronize before beginning the edit.

predicted frames In motion compensation compression, the case where a new frame is based on the previous (or sometimes the future) frame. Because frames depend on one another, playback or editing cannot be started at a predicted frame. See interframe coding.

primary colors In a color video system, the color names for the three different color response curves of the camera channels. The most common system of taking curves is RGB.

print-through In a magnetic tape, the property where the recorded information in one layer of tape rolled on a reel may affect the recordings on adjacent layers, causing interference.

production In the style of program creation where all material is captured on tape or disk before beginning program editing and assembly, the shooting and capture phase. See postproduction.

progressive scanning A scanning system where all lines of the raster are scanned in each vertical scan. See interlaced scanning.

pulse code modulation (PCM) In digital transmission, direct encoding of audio or video sample values.

quadruplex The earliest videotape recording system, developed by Ampex. It used four heads on a rotating wheel to record transverse tracks on a wide tape.

quantizing In analog-to-digital conversion, the process of assigning a digital value to each sample.

RAID See redundant array of inexpensive disks.

raster In a scanned video system, the total pattern of lines produced by the scanning motions. Most rasters consist of a rectangular pattern of lines, fields, and frames.

Rec. BT.601 A standard developed by ITU-R for component digital video. See International Telecommunications Union.

recordable CD (CD-R) A version of the CD-ROM that can be recorded in a PC. A special drive is used that can write once on the CD surface. Another version is called the rewritable CD, which can also erase and rewrite recordings. Recordable CDs and rewritable CDs can be played by most standard CD-ROM drives.

recording density In recording systems, the number of bits stored in a given area of the medium. For example, 1 Mb/cm^2. More efficient recording systems have higher recording density.

Red Book The standards document for the audio-CD system.

redundant array of inexpensive disks (RAID) A mass storage technique using multiple hard disks to provide greater throughput and backup capability.

Reed-Solomon coding An error protection method involving algorithms for generating error protection codes for specific-sized blocks of data to give a specified degree of error correction capability to each block.

relative permeability See permeability.

remanent magnetization The magnetization remaining in a magnetic material when all magnetizing force is removed. Defined by the parameter of remanence.

requantizing In a sampled digital system, when the number of bits per sample is reduced, requantizing makes sure that as much information as possible is kept in the remaining bits.

resolution In a video system, the ability to reproduce fine detail. Video resolution is measured in TV lines of resolution. See horizontal resolution, vertical resolution.

rewritable CD (CD-RW) See recordable CD.

RLE See run length encoding.

RGB Acronym for red-green-blue, the three most common camera primary colors.

routing switcher In a studio system, a switching system that connects cameras, VCRs, and other units to the users who will control them.

run length encoding (RLL) In digital systems, an encoding scheme that codes repeated values in the data stream as a count number and a value.

S-Video An analog video interface and connector standard that sends separate luminance and chrominance values.

sample and hold A sampling process that holds its value from one sample to the next, thus producing a stepped waveform rather than a series of pulses.

sampling The process of representing an analog signal as a series of equally spaced pulses whose amplitudes equal the value of the analog signal at their time of occurrence.

sampling rate The rate of occurrence of sampling.

saturation In a magnetic circuit, the property that causes limiting of magnetization regardless how large the magnetizing force becomes.

scan conversion In a video system, the process of converting video data from one scanning standard to another.

scanning In a video system, the process of reading across and down an image to convert it into a time-sequential video signal. Most scanning systems use a rectangular pattern called a raster, composed of lines, fields, and frames.

SDTV See standard-definition television.

SECAM See Sequential Couleur Avec Mémoire.

sector In disk recording, a portion of a track that is associated with a specified amount of data. Sectors may be defined as a certain angle of disk rotation in CAV systems, or as a certain track length in CLV systems.

separation loss In magnetic recording, the loss of signal caused by any nonmagnetic separation between the head gap and the recording layer.

Sequential Couleur Avec Mémoire (SECAM) The composite color television system developed in France and used there and in the countries of the former Soviet Union. SECAM uses a baseband luminance signal and two frequency-modulated subcarriers that carry the color-difference information. See color-difference signals.

serial process In an electronic system, where processes occur sequentially in a single signal path.

server A recording system whose purpose is to store information and provide it on demand to one or more users.

servomechanism An electromechanical system that automatically controls a mechanical motion by means of electrical feedback. Typical servomechanisms in recording control the rotation of a disk or scanner, movement of a head, and so on. Abbreviation: servo.

shuttle mode In videotape recording, a mode of tape deck operation where the tape speed is continuously variable from reverse to fast forward. See picture-in-shuttle.

signal processing In an electronic system, processes that modify, adjust, or combine signals.

signal-to-noise ratio (SNR) In an electronic system, the ratio between the normal signal level and the noise present with no signal or with a uniform signal. In video systems, SNR is expressed in decibels and is the ratio between peak-to-peak black-to-white video and the rms noise level.

SMPTE See Society of Motion Picture and Television Engineers.

SNR See signal-to-noise ratio.

Society of Motion Picture and Television Engineers (SMPTE) An organization of engineers that develops standards in the fields of audio, video, recording, and motion pictures.

spatial compression Video compression techniques that exploit the redundancy between pixels and lines to accomplish compression.

special effects In postproduction, the use of dynamic transitions between video signals, such as wipes, dissolves, rotations, page-turnings, and so forth.

standard-definition television (SDTV) Television systems having the same scanning standards as the existing analog television systems, such as NTSC, PAL, or SECAM. The term is used loosely and sometimes also refers to modifications of the existing standards, such as using progressive scanning or 16:9 aspect ratio with 525 lines, or using the 640 × 480 pixel computer standard for television.

statistical coding A digital encoding method that exploits data statistics, such as frequency of occurrence, patterns, and so on. See Huffman coding.

subsampling In a sampling system, the use of a divided-down sampling rate for one part of the system compared with another. For example, in a component video system of the 4:2:2 type, the color-difference components are sampled at one-half the luminance sampling rate.

subtractive color A color reproduction system where display is accomplished by placing colored dyes on a white reflective surface. Uses of this are in color printing or color print photography. The primaries for subtractive color are usually cyan, magenta, and yellow, although a fourth color (black) may also be added to improve black rendition, especially in color printing. That system is called CMYK.

symbol In a digital transmission system, the basic unit of data carried at one moment by the channel signal. A single symbol may carry 1 or more bits.

sync generator A circuit that generates the horizontal and vertical timing signals to control the scanning of multiple video devices so they will be synchronous.

tachometer A device mechanically connected to a rotating member to generate electrical signals representative of the member's rotation. Tachometers are often used in servomechanisms to provide feedback of the angular position or speed of a rotating shaft.

tape A substrate designed in the form of a long strip of flexible material that can be rolled onto reels for storage. The most common tape in recording is magnetic tape.

tape deck In a recording system, the mechanism that handles unrolling of tape from the supply reel, passing it over the magnetic heads or the helical scanner, and rolling it up on a takeup reel. Other tape deck functions may include cassette loading and tape rewinding. Sometimes called a tape transport.

TBC See time base corrector.

TDM See time-division multiplexing.

television A video system specifically designed for distribution of pictures and sound to a mass audience. Distribution may be by means of broadcasting, either terrestrial or satellite, or by cable.

temporal compression Video compression techniques that exploit the redundancy between successive frames of a video sequence. See motion compensation.

tension arms Components of a tape deck that sense the tension in the tape, usually by wrapping the tape around a post that is spring-mounted on an arm. The position of the arm is proportional to the tape tension. Tension arms provide the feedback signal for servomechanisms that control tape tension in a tape deck.

test tape A tape recording of test signals that is certified to meet a particular standard. Satisfactory playing of a test tape provides partial assurance that a given recorder meets its standard.

thickness loss In magnetic recording, a signal loss associated with the thickness of the magnetic recording layer.

thin-film heads Magnetic heads fabricated by the thin-film processes generally used for IC fabrication.

threading In a tape deck, the process of initially passing the tape through the tape path. Threading may be manual or automatic, although most modern tape decks use cassette tapes with automatic threading.

three-spot method In optical recording, a tracking servomechanism approach that splits the laser beam into three spots that straddle the tracks.

time base corrector (TBC) A circuit or module that corrects time base errors resulting from nonuniform mechanical motions in a recording mechanism.

time base error Frequency or phase instabilities in the smooth flow of information from a recording device.

time code In recording systems for audio or video, a special code recorded on the same medium along with the audio or video information for the purpose of uniquely identifying video frames and the corresponding audio. This is necessary for precise editing of audio and video.

time-division multiplexing (TDM) In an electronic system, the combining of multiple signals in the same channel by assigning them segments of time. TDM is commonly used for digital multiplexing.

tracking The process of controlling head motion to pass accurately over recorded tracks for playback. Optical recorders also often use tracking during recording to assure that the recorded tracks meet their desired standard.

tracks The pattern recorded on a recording medium to provide a single recording channel. Tracks on tape may be longitudinal (along the tape) or helical (at an angle to the tape edge). Tracks on disk are either concentric circles or spirals.

transitions In editing, the visual effect of the change from one camera shot to the next. Most systems provide a variety of effects including cuts, dissolves, fades, wipes, and other dynamic effects.

transparent The property of a system to not noticeably change signals passing through it.

trichromatic color A color reproduction system having three independent channels with different spectral responses to light. The three spectral responses are referred to by their dominant colors, such as red, green, and blue; these are the primary colors for the system.

TV See television.

TV lines of resolution (TVL) The measure of resolution in a video system. TVL is defined as the number of equally spaced white and black lines in a distance equal to the picture height.

TVL See TV lines of resolution.

Type C The SMPTE designation for the 1-in analog helical scan tape system widely used in broadcasting.

VBI See vertical blanking interval.

vertical blanking interval (VBI) In an analog TV signal, the period of time allowed for the receiver scanning beam to retrace back to the start of the next field.

vertical interval time code (VITC) A time code system where the time data is included in the vertical blanking interval of a TV signal.

vertical resolution Picture resolution in the vertical direction; that is, observed on horizontally oriented lines or patterns.

VHS See Video Home System.

video The electronic representation of pictures, either stationary or moving.

video compression The application of data compression technology to video signals.

Video Home System (VHS) A magnetic tape recording system designed for home use. VHS uses 1/2-in tape in cassettes and is widely used for distribution of prerecorded video to the home.

video on demand (VOD) A video distribution system where users can choose the video they want to watch at any particular time. This is generally done by means of a video server.

video recorder A recording device designed to record video signals.

video system A system for capturing, recording, processing, or reproduction of video.

viewing ratio In viewing a video display, the ratio between the distance from viewer to display and the picture height. The closer the viewer sits to a given display, the smaller the viewing ratio and the greater resolution he or she will be able to perceive on the display.

VITC See vertical interval time code.

VOD See video on demand.

wavelength In any wave phenomenon, the ratio between the velocity of propagation and the wave frequency.

white balance In a video system, the condition where perceived whites in the scene are reproduced as the specified white condition for the system. For example, most RGB systems specify white as equal R, G, and B signals.

wipes In editing, transition effects where one or more lines of demarcation move across the picture to create the transition from one shot to the next.

writing speed In a recording system, the speed at which tracks pass by the recording or replaying head.

YC_RC_B format A component video format consisting of luminance (Y) and two color-difference components based on $R - Y$ and $B - Y$. See color-difference signals.

Yellow Book The standard document for the CD-ROM.

Bibliography

GENERAL VIDEO AND TELEVISION

Baron, S. N. (ed.), *Implementing HDTV: Television and Film Applications*, SMPTE, White Plains, NY, 1996.

Benson, K. B., and Whitaker, J., *Television Engineering Handbook*, McGraw-Hill, New York, 1992.

Fisher, D. E., and Fisher, M. J., *TUBE: The Invention of Television*, Counterpoint, Washington, DC, 1996.

Inglis, A. F., and Luther, A. C., *Video Engineering,* 2nd ed., McGraw-Hill, New York, 1996.

Zettl, H., *Television Production Handbook,* 6th ed., Wadsworth Publishing, Belmont CA, 1997.

DIGITAL AND COMPUTER TECHNOLOGY

Dorf, R. C., *Electrical Engineering Handbook*, CRC Press, Boca Raton, FL, 1993.

Luther, A. C., *Principles of Digital Audio and Video*, Artech House, Norwood, MA, 1997.

Jayant, N. S., and Noll, P., *Digital Coding of Waveforms*, Prentice-Hall, Englewood Cliffs, NJ, 1984.

Negroponte, N., *Being Digital*, Knopf, New York, 1995.

Rzeszewski, T. S. (ed.), *Digital Video: Concepts and Applications Across Industries*, IEEE Press, New York, 1995.

Watkinson, J., *The Art of Digital Video,* 2nd ed., Focal Press, Oxford, 1994.

RECORDING TECHNOLOGY

Jorgenson, F., *The Complete Handbook of Magnetic Recording,* 4th ed., McGraw-Hill, New York, 1996.

Mee, C. D., and Daniel, E. D., *Magnetic Storage Handbook,* 2nd ed., McGraw-Hill, New York, 1996.

Mee, C. D., and Daniel, E. D., *Magnetic Recording* (three volumes), McGraw-Hill, New York, 1987.

Watkinson, J., *The Art of Data Recording*, Focal Press, Oxford, 1994.

Watkinson, J., *The Digital Video Tape Recorder*, Focal Press, Oxford, 1994.

OPTICAL TECHNOLOGY

Meyer-Arendt, J. R., *Introduction to Classical and Modern Optics*, 4th ed., Prentice-Hall, Englewood Cliffs, 1995.

Nadeau, M., *Byte Guide to CD-ROM*, 2nd ed., Osborne McGraw-Hill, Berkeley, 1995.

Purcell, L., and Martin, D., *The Complete Recordable-CD Guide*, Sybex, San Francisco, 1997.

Taylor, J., *DVD Demystified*, McGraw-Hill, New York, 1998.

Williams, E. W., *The CD-ROM and Optical Disc Recording Systems*, Oxford University Press, Oxford, 1996.

AUDIO

Benson, K. B., *Audio Engineering Handbook*, McGraw-Hill, New York, 1988.

Pohlmann, K. C., *Principles of Digital Audio,* 3rd ed., McGraw-Hill, New York, 1995.

EDITING AND POSTPRODUCTION

Anderson, G., *Video Editing and Postproduction: A Professional Guide,* 3rd ed., Focal Press, Boston, 1993.

TELECOMMUNICATIONS

Agnew, P. W., and Kellerman, A. S., *Distributed Multimedia: Technologies, Applications, and Opportunities in the Digital Information Industry*, ACM Press, New York, 1996.

Gibson, J. D. (ed.), *The Communications Handbook*, CRC Press, Boca Raton, FL, 1997.

Inglis, A. F., and Luther, A. C., *Satellite Technology: An Introduction,* 2nd ed., Focal Press, Boston, 1997.

Lu, G., *Communication and Computing for Distributed Multimedia Systems*, Artech House, Norwood, MA, 1996.

Minoli, D., and Keinath, R., *Distributed Multimedia Through Broadband Communications Services*, Artech House, Norwood, MA, 1994.

Ohta, N., *Packet Video: Modeling and Signal Processing*, Artech House, Norwood, MA, 1994.

Riley, M. J., and Richardson, I. E. G., *Digital Video Communications*, Artech House, Norwood, MA, 1997.

Schaphorst, R., *Videoconferencing and Videotelephony: Technology and Standards*, Artech House, Norwood, MA, 1996.

STANDARDS

Advanced Television Systems Committee (ATSC), http://www.atsc.org

Digital Video Broadcasting (DVB), http://www.dvb.org

Institute of Electrical and Electronics Engineers (IEEE), http://www.stdsbbs.ieee.org

International Standards Organization (ISO), http://www.iso.ch

International Telecommunications Union (ITU), http://www.itu.ch

Society of Motion Picture and Television Engineers (SMPTE), White Plains, NY, http://www.smpte.org

PERIODICALS

AV Video, published monthly by Montage Publishing, Inc., 701 Westchester Ave., White Plains, NY 10604.

Broadcast Engineering, published monthly by PRIMEDIA Intertec, 9800 Metcalf, Overland Park, KS 66212-2215.

Byte, published monthly by the McGraw-Hill Companies, Inc., P.O. Box 552, Hightstown, NJ 08520.

Digital Video, published monthly by Miller-Freeman, Inc., 600 Harrison St., San Francisco, CA 94107.

IEEE Transactions on Consumer Electronics, IEEE Operations Center, 445 Hoes Lane, P.O. Box 1331, Piscataway, NJ 08855-1331.

IEEE Transactions on Broadcasting, IEEE Operations Center, 445 Hoes Lane, P.O. Box 1331, Piscataway, NJ 08855-1331.

Millimeter, published monthly by Intertec Publishing Corporation, 5 Penn Plaza, 13th Floor, New York, NY 10001.

New Media, published monthly by HyperMedia Communications, P.O. Box 3039, Northbrook, IL 60065-3039.

SMPTE Journal, published monthly by the Society of Motion Picture and Television Engineers, White Plains, NY, http://www.smpte.org

Videography, published monthly by Miller-Freeman PSN, Inc., 460 Park Avenue South, New York, NY 10016.

Videomaker, published monthly by Videomaker, Inc., P.O. Box 4591, Chico, CA 95927.

OTHER WEB SITES

http://www.amazon.com (bookseller)

http://www.antonbauer.com (batteries)

http://www.belden.com (wire and cable)

http://www.hitachi.com (recorders and camcorders)

http://www.jvc-america.com (recorders and camcorders)

http://www.panasonic.com (recorders and camcorders)

http://www.sarnoff.com (video research)

http://www.sel.sony.com (recorders and camcorders)

http://www.snellwilcox.com (standards conversion, synchronizers)

http://www.tek.com (measurement)

Index

The Artech House Digital Audio and Video Library

For further information on these and other Artech House titles, including previously considered out-of-print books now available through our In-Print-Forever™ (IPF™) program, contact:

Artech House
685 Canton Street
Norwood, MA 02062
781-769-9750
Fax: 781-769-6334
Telex: 951-659
e-mail: artech@artech-house.com

Artech House
46 Gillingham Street
London SW1V 1AH England
+44 (0) 171-973-8077
Fax: +44 (0) 171-630-0166
Telex: 951-659
e-mail: artech-uk@artech-house.com

Find us on the World Wide Web at:
www.artech-house.com